HODDER
mathematics
INTERMEDIATE
SECOND EDITION

Series editor: Roger Porkess

Catherine Berry
Pat Bryden
Diana Cowey
Dave Faulkner
John Spencer

Hodder & Stoughton
A MEMBER OF THE HODDER HEADLINE GROUP

Acknowledgements

The authors and publishers would like to thank the following companies, institutions and individuals who have given permission to reproduce copyright material: Office for National Statistics, Wales Tourist Board.

The publishers will be happy to make arrangements with any copyright holders whom it has not been possible to contact.

Illustrations were drawn by Bill Donohoe and Mark Walker of Ian Foulis and Associates, Maggie Brand, Tom Cross and Joe McEwan.

Photos supplied by Wales Tourist Board Photo Library (page 60), Patsy Lynch/Corbis (page 130 left), Frank Lee/Corbis (page 130 right).

Page design and cover design by Lynda King.

Orders: please contact Bookpoint Ltd, 130 Milton Park, Abingdon, Oxon OX14 4SB. Telephone: (44) 01235 827720, Fax: (44) 01235 400454. Lines are open from 9.00 – 6.00, Monday to Saturday, with a 24 hour message answering service.

British Library Cataloguing in Publication Data

A catalogue record for this title is available from The British Library

ISBN 0 340 803 738

First published 1998
Second edition 2001
Impression number   10 9 8 7 6 5 4 3 2
Year   2006 2005 2004 2003 2002

Copyright © 1998, 2001 Catherine Berry, Pat Bryden, Diana Cowey, Dave Faulkner, Roger Porkess, John Spencer

All rights reserved. No part of this publication may be reproduced or transmitted in any form or by any means, electronic or mechanical, including photocopy, recording, or any information storage and retrieval system, without permission in writing from the publisher or under licence from the Copyright Licensing Agency Limited. Further details of such licences (for reprographic reproduction) may be obtained from the Copyright Licensing Agency Limited, of 90 Tottenham Court Road, London W1P 9HE.

Cover photo from Science Photo Library

Typeset by Pantek Arts Ltd, Maidstone, Kent, ME14 1NY

Printed in Italy for Hodder & Stoughton Educational, a division of Hodder Headline Plc, 338 Euston Road, London NW1 3BH by Printer Trento.

Introduction to the second edition

This is the first of two textbooks covering Intermediate Tier GCSE. Students following a two-year course would expect to take one year on each book. The books cover the requirements of Intermediate Tier GCSE and so are suitable for use with any specification. The division of material between them ensures both coverage of the modules within the MEI GCSE specification and a balanced teaching curriculum over the two years. This book also covers the mathematics requirements of GNVQ Application of Number at Level 3.

This is the second edition of this book. It has been adapted to take account of new GCSE criteria. These apply to courses with first teaching in September 2001 and first certification in summer 2003. The changes are considerable and this new edition incorporates all of them.

The book is divided into 16 chapters, forming a logical progression through the material (some teachers may however wish to vary this order). Each chapter is divided into a number of double-page spreads, designed to be teaching units. The material to be taught is covered on the left-hand page; the right-hand pages consist entirely of work for the students to do. Each chapter ends with a mixed exercise covering all of its content. Further exercise sheets and tests are provided in the Teacher's Resource.

The instruction (i.e. left-hand) pages have been designed to help teachers engage their students in whole class discussion. The symbol is used to indicate a Discussion Point; teachers should see it as an invitation.

Most of the right-hand pages end with a practical activity. These are suitable for both GCSE and GNVQ students; some can be used for portfolio tasks. Advice on these is available in the Teacher's Resource and, where relevant, raw data is also supplied. Most students will not do all of the activities (they can be quite time-consuming) but the authors think it is important that they do as many of them as possible; they connect the mathematics classroom to the outside world and to other subjects.

Where knowledge is assumed, this is stated at the start of the chapter. There is a general expectation that students will know the content of Foundation Tier GCSE. Questions indicated with a calculator icon need to be answered with a calculator. The 'no calculator' icon indicates that a calculator should definitely not be used. Many questions have neither icon and these require a sensible judgement. Students should do as many of these as possible without a calculator in order to practise for the non-calculator GCSE questions. However, they also need to work through plenty of questions and using a calculator often allows them to work faster.

Although students are to be encouraged to use I.T., particularly spreadsheets, specific guidance is limited to the Teacher's Resource. Otherwise, the book would have been based on one particular package to the frustration of those using all the others.

The authors would like to thank all those who helped in preparing this book, particularly Chris Curtis for his advice on early versions of the manuscript, and Karen Eccles who typed much of the first edition.

Contents

Information pages 2

Chapter One: Whole numbers and decimals 4
    About numbers 4
    Standard form 6
    Decimals 8
    Imperial and metric units 10
    Prime factorisation 12

Chapter Two: Shapes and angles 16
    Reminder 16
    Quadrilaterals 18

Chapter Three: Starting algebra 22
    Writing things down 22
    The language of algebra 24
    Substituting into a formula ... 26
    Using brackets 28
    Adding and subtracting with negative numbers 30

Chapter Four: Fractions and percentages 34
    Reminder 34
    Using fractions 36
    From fractions to percentages ... 38
    Using percentages 40
    Making comparisons 42

Chapter Five: Area and volume 46
    Parallelograms and trapezia 46
    Circumference and area of a circle ... 48
    Volume of a prism 50
    Surface area of a prism 52

    More about prisms 54
    Using dimensions 56

Chapter Six: Using symbols 60
    Being brief 60
    Using negative numbers 62
    Simplifying expressions with negative numbers 64

Chapter Seven: Data handling 68
    Pie charts 68
    Stem-and-leaf diagrams 70
    Moving averages 72
    Bivariate data 74
    Line of best fit 76

Chapter Eight: Graphs 80
    Looking at graphs 80
    Gradient and intercepts 82
    Obtaining information 84
    Travel graphs 86
    Curved graphs 88

Chapter Nine: Ratio and proportion 92
    Revision exercise 92
    Using ratio 94
    Unitary method 96
    Changing money 98
    Distance, speed and time 100

Chapter Ten: Grouped data ... 104
    Grouping continuous data ... 104
    Grouping rounded data 106
    Frequency polygons 108
    Mean, median and mode of grouped data 110

Contents

Cumulative frequency 112
Quartiles 114
Box-and-whisker diagrams
(boxplots) 116

Chapter Eleven: Equations 120
Using equations 120
Solving equations 122
Equations with fractions 124
Changing the subject of a
formula 126

Chapter Twelve: Approximations 130
Decimal places 130
Significant figures 132
Estimating costs 134
Using your calculator 136
Using fractions and percentages 138
Errors 140

Chapter Thirteen: Practical statistics 144
Revision exercise 144
Reminder 146
Which average? 148
Making comparisons 150
Conducting a survey 152
Writing your report 154

Chapter Fourteen: Finding lengths 156
Scale drawings 156
Using bearings 158
Pythagoras' rule 160
Finding one of the shorter sides 162

Chapter Fifteen: Money 166
Simple interest 166
Compound interest 168
Bills 170
Hidden extras 172
Profit and loss 174
Repeated changes 176

Chapter Sixteen: Probability ... 180
Working out probability 180
Estimating probability 182

Answers 186

Index 202

Information

How to use this book

 This symbol next to a question means you need to use your calculator.

 This symbol next to a question means you are not allowed to use your calculator.

 This symbol means you will need to think carefully about a point and may want to discuss it.

Triangles

An **equilateral** triangle has 3 equal sides.

An **isosceles** triangle has 2 equal sides.

A **scalene** triangle has no equal sides.

A **right-angled** triangle has 1 right angle.

An **acuted-angled** triangle has 3 acute angles.

An **obtuse-angled** triangle has 1 obtuse angle.

Area of a triangle = $\frac{1}{2}$ × base × height

Quadrilaterals

  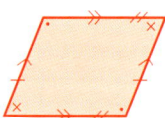 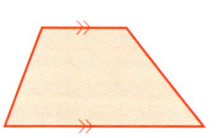 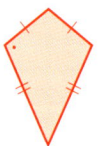 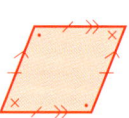

square   rectangle   parallelogram   trapezium   kite   rhombus

Area of a parallelogram = base × vertical height

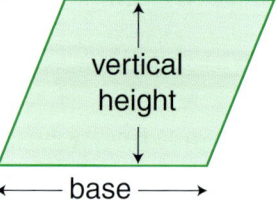

Area of a trapezium = $\frac{1}{2}(a+b)h$

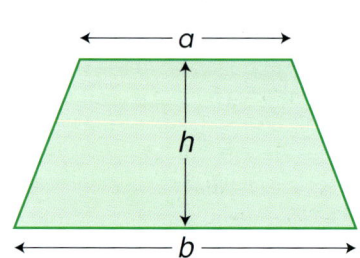

Circles

Circumference of circle = π × diameter
 = 2 × π × radius

Area of circle = π × (radius)^2

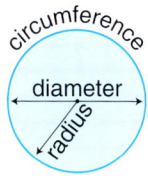
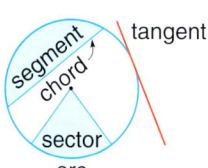

Solid figures

Volume of cuboid = length × width × height

Volume of prism = area of cross section × length

Volume of cylinder = πr^2 × length

Trigonometry

$\sin \theta = \dfrac{\text{opposite}}{\text{hypotenuse}} = \dfrac{y}{h}$

$\cos \theta = \dfrac{\text{adjacent}}{\text{hypotenuse}} = \dfrac{x}{h}$

$\tan \theta = \dfrac{\text{opposite}}{\text{adjacent}} = \dfrac{y}{x}$

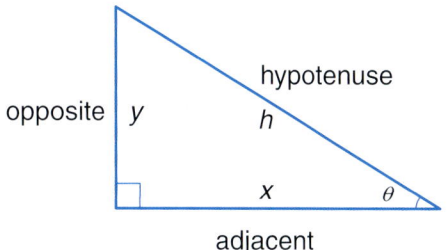

Pythagoras' rule: $x^2 + y^2 = h^2$

Units

Metric system

Length

k   1 kilometre = 10^3 metres = 1000 metres

h   1 hectometre = 10^2 metres = 100 metres

da   1 decametre = 10^1 metres = 10 metres

d   1 decimetre = 10^{-1} metres = $\dfrac{1}{10}$ metre

c   1 centimetre = 10^{-2} metres = $\dfrac{1}{100}$ metre : 100 centimetres = 1 metre

m   1 millimetre = 10^{-3} metres = $\dfrac{1}{1000}$ metre : 1000 millimetres = 1 metre

The units for mass and capacity follow the same pattern. Notice that:
1 kilogram = 1000 grams   1 litre = 1000 millilitres

Notice also that: 1 tonne = 1000 kg

Imperial

12 inches = 1 foot   16 ounces = 1 pound
3 feet = 1 yard   14 pounds = 1 stone
1760 yards = 1 mile   8 stones = 1 hundredweight (cwt)
                                    20 cwt = 1 ton

One

Whole numbers and decimals

> **Before you start this chapter you should be able to**
>
> ★ write a number in figures
> ★ add and subtract decimals
> ★ convert simple fractions into decimals and decimals into fractions
> ★ read decimal scales
> ★ convert from one unit to another (e.g. cm to m)
> ★ do calculations involving time
> ★ find multiples, factors, primes, squares, square roots, cubes and cube roots and powers.

About numbers

Numbers are an important part of everyday life. You need to be able to work with them confidently. Here are some terms you will find helpful.

Multiple   The multiples of 4 are 4, 8, 12, 16, ...
Factor   The factors of 20 are 1, 2, 4, 5, 10 and 20
               A factor may also be called a **divisor**.
Prime   17 is prime; its only factors are 1 and itself, 17
Power 2^4 is called '2 to the power 4'. It means $2 \times 2 \times 2 \times 2$ or 16
Index form When 16 is written as 2^4 it is in index form. The index is 4 in this case. The plural of index is indices.

Exercise

1 a) Round the following numbers to the nearest 10.
   (i) 73   (ii) 49   (iii) 262   (iv) 488

b) Round the following numbers to the nearest 100.
   (i) 2149   (ii) 3175   (iii) 4912   (iv) 7979

c) Round the following numbers to the nearest 1000.
   (i) 2433   (ii) 7612   (iii) 59 147   (iv) 103 103

2 Express

a) 30 minutes as a fraction of 4 hours

b) 55p as a fraction of £5

c) 325 m as a fraction of 6.5 km

d) 75 g as a fraction of 3 kg.

1: Whole numbers and decimals

Exercise

3 Write in figures
   a) four thousand six hundred and three
   b) twenty thousand and eighteen
   c) three and one tenth
   d) two and nineteen thousandths.

4 Write each of these fractions as a decimal.
   a) $\frac{3}{10}$ b) $\frac{87}{100}$ c) $5\frac{7}{10}$
   d) $2\frac{1}{100}$ e) $4\frac{113}{1000}$ f) $\frac{41}{1000}$

5 Write each of these decimals as a fraction in its simplest form.
   a) 0.2   b) 0.91   c) 4.9   d) 1.87   e) 3.25   f) 0.005

6 Write down each of these readings.

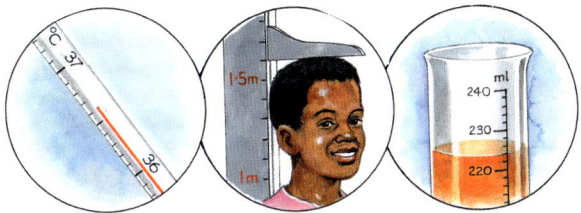

7 Work out each of these.
   a) 7.3 + 8
   b) 6.7 + 4.03 + 0.3
   c) £5 − £3.65
   d) 22.4 − 3.97
   e) 14.29 − (3.2 + 6.854)
   f) £3.27 + 45p

8 Convert
   a) 5 cm into mm
   b) 70 cl into l
   c) 4 kg into g
   d) 20 ozs into lbs and oz
   e) 8 mm into cm
   f) 0.5 l into ml
   g) 1500 m into km
   h) 5 ft into inches
   i) 250 g into kg
   j) 2 l into cl
   k) 5 gallons into pints
   l) 3 km into m.

9 Write down six multiples of 7.

10 List all the factors of
   a) 15   b) 24   c) 40   d) 45   e) 60   f) 72   g) 100   h) 144

11 Work out which of these numbers are primes.
   a) 17   b) 26   c) 31   d) 39   e) 73   f) 91

12 Work out
   a) 5^2 b) 6^3 c) $\sqrt{400}$ d) 30^2 e) $\sqrt[3]{27}$
   f) 2^{10} g) $\sqrt[3]{125}$ h) 10^6 i) $\sqrt{9} \times \sqrt{9}$ j) $\sqrt{3} \times \sqrt{3}$

13 Write these numbers in order of size, smallest first
   a) 6 million, 7×10^5, 30 000 000, 25K
   b) 5 hundredths, 8×10^{-2}, 0.009

Investigate how your calculator displays very large and very small numbers.

Check that you can use your calculator to find cubes and cube roots.

5

1: Whole numbers and decimals

Standard form

Large numbers

The speed of light is 300 000 000 m/s.

Large numbers like this are rather untidy.

300 000 000 is the same as

$$3 \times 100\ 000\ 000$$

or $3 \times 10 \times 10 \times 10 \times 10 \times 10 \times 10 \times 10 \times 10$

This can be written as 3×10^8.

This is an example of **standard form**. The leading number, 3, is between 1 and 10.

The speed of sound in standard form is 3.3×10^2 m/s.

What is 3.3×10^2 in decimal form?

$$3.3 \times 10^2 = 3.3 \times 10 \times 10 = 330$$

Small numbers

The wavelength of yellow light is 0.000 000 6 metres.

Small numbers like this are also untidy.

$$0.000\ 000\ 6 = \frac{6}{10\ 000\ 000} \text{ or } \frac{6}{10 \times 10 \times 10 \times 10 \times 10 \times 10 \times 10}$$

This can be written as $\frac{6}{10^7}$ or 6×10^{-7}. ← *This is in standard form.*

The wavelength of mercury green light is 5.4×10^{-7} m.

$$5.4 \times 10^{-7} = \frac{5.4}{10^7} = \frac{5.4}{10 \times 10 \times 10 \times 10 \times 10 \times 10 \times 10} = 0.000\ 000\ 54$$

The wavelength of mercury green light is 0.000 000 54 m.

This number, 0.000 000 54, can also be written as 54×10^{-8}.

This is not in standard form because the leading number, 54, is not between 1 and 10. You can convert it like this

$$54 \times 10^{-8} = 5.4 \times 10 \times 10^{-8} \quad \text{(because } 54 = 5.4 \times 10\text{)}$$
$$= 5.4 \times 10^{-7} \quad \text{(because } 10 \times 10^{-8} = 10^{-7}\text{)}$$

1: Whole numbers and decimals

1 These numbers are in standard form.

Write them out in full.

a) 6×10^2 b) 3×10^4 c) 7×10^{-3} d) 4×10^{-5}
e) 4.5×10^6 f) 5.4×10^{-3} g) 9.4×10^3 h) 8.75×10^{-4}
i) 1.6×10^{-2} j) 2.75×10^6 k) 8.3×10^{-5} l) 1.05×10^4
m) 7.3×10^3 n) 8×10^{-9} o) 4×10^{-1} p) 8.25×10^{10}

2 Write these numbers in standard form.

a) 4000   b) 800 000   c) 0.003   d) 0.0009
e) 26 000   f) 0.025   g) 7 500 000   h) 0.000 037
i) 810   j) 0.005 43   k) 0.93   l) 64 000
m) 0.016   n) 147 000 000   o) 0.507   p) 9040

3 Alison is using this spreadsheet for her science assignment.

	D	E	F
1	637 983	5E–03	4.38E+09
2	7.25E–04	9.42E+08	694
3	4.6E+12	83 926	7.5E–06

Large and small numbers are displayed using an E.

So in D2, 7.25E–04 means 7.25×10^{-4}.

Work out the value of the entry in

a) E1   b) D3   c) E2   d) F3

4 This table shows the approximate diameter of each planet.

Planet	Diameter (m)
Mercury	4.88×10^6
Venus	1.21×10^7
Earth	1.28×10^7
Mars	6.79×10^6
Jupiter	1.44×10^8
Saturn	1.21×10^8
Uranus	5.08×10^7
Neptune	4.95×10^7
Pluto	2.30×10^6

Arrange them in order of size.

Which of these numbers are the same?
1.2×10^{-1}, 1 200 000, $\frac{3}{25}$, 0.12, 1.2×10^6, one million two hundred thousand

The population of Sweden is about 8 800 000 or 8.8×10^6.

Find out the approximate population of nine other countries.

Present your results in a table, showing the populations in both standard form and as ordinary numbers.

7

1: Whole numbers and decimals

Decimals

Paul is leading a party of 12 people (including himself) on a hiking weekend. They stay one night at a hostel that charges £11.35 per person.

How much does it cost the group?

The cost is 12 × £11.35. Paul works it out like this:

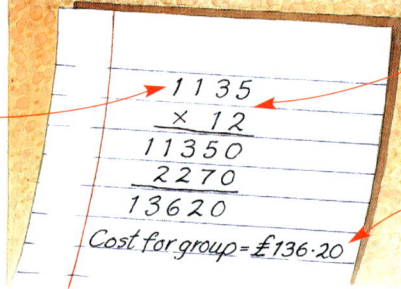

He makes the 11.35 into a whole number by moving the decimal point 2 places to the right…

…then he multiplies the whole numbers

He moves the decimal point 2 places to the left to get his final answer

 On the second night they stay at a hostel charging £12.15 per person. How much does it cost the group?

Why does Paul's method work?

Paul and Angela share the driving. They drive the minibus 186 miles during the weekend. It does 6.2 miles per litre of petrol.
How many litres of petrol does it use?

Paul works it out like this:

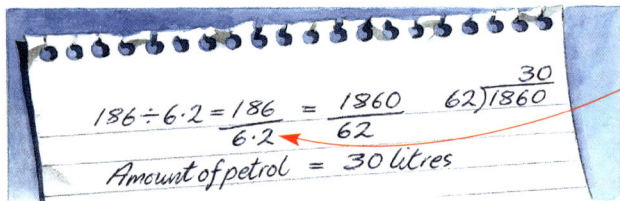

This must be a whole number before you divide. The top and bottom of the fraction have been multiplied by 10

The minibus uses 30 litres of petrol.

Quick methods

Paul works out the cost of 30 litres of petrol at 71.9 pence per litre.

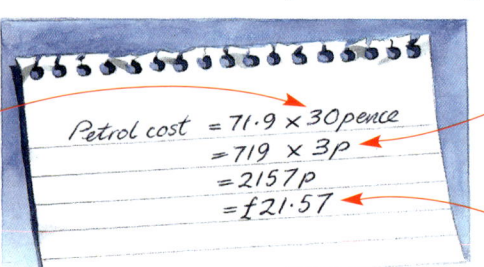

Multiplying by 30 is the same as × 10 and then × 3

Multiplying by 10 moves the decimal point one place to the right. 71.9 becomes 719

2157 pence is £21.57

 The minibus drivers, Paul and Angela, travel free. The other 10 people share the cost of petrol. How much does each pay?

1: Whole numbers and decimals

1
a) 3×1.45 b) 3.2×1.5 c) 0.6×5 d) 6×0.568
e) 1.2×3.5 f) 4.16×3 g) 0.2×5.2 h) 2.5^2

2
a) $6 \div 0.3$ b) $16.8 \div 1.2$ c) $15.4 \div 0.35$ d) $5 \div 0.8$
e) $4 \div 0.5$ f) $7.2 \div 2$ g) $5.6 \div 1.2$ h) $0.27 \div 0.9$

3

Malta	27 °C
Cyprus	29 °C
Tunisia	36 °C

You can convert temperatures from Celsius to Fahrenheit by multiplying by 1.8 and adding 32. Convert each of these, giving your answer to the nearest degree.

4 Harriet buys 1.5 kg of bananas for 72 pence, and 2.5 kg of potatoes for 65 pence.

Work out the cost per kg of
a) bananas
b) potatoes.

5 Hana's car travels 7.2 miles on one litre of petrol.

Work out, to the nearest litre, how much petrol she needs for a journey of

a) 115 miles b) 295 miles

c) Petrol costs 71.5p per litre. Work out the cost of petrol to the nearest 10p for each journey.

6 A nursing home has 20 residents each paying £400 a week.

a) What is the total income per week?

b) What is the total income per year?

7 Work out the total cost of this stationery order.

Item	Unit cost (£)	Quantity	Cost (£)
Pens	0.16	300	
Pads of paper	0.80	50	
Files	1.20	200	
Envelopes	0.05	4000	
		Total	

Try out the following mental arithmetic test on your friends.

1. 30×10 2. $600 \div 20$
3. 0.1×50 4. 40×700
5. $1200 \div 30$ 6. $\frac{1}{10} \times 900$
7. 250×40 8. 0.01×600
9. $8000 \div 20$ 10. $10 \times 20 \times 50$

Now make up a test of your own to try out on your friends.

1: Whole numbers and decimals

Imperial and metric units

It is useful to be able to make quick conversions between different types of units. These conversions are worth remembering.

Length

1 inch is about 2.54 cm

39 inches is about 1 m

1 foot is about 30.5 cm

$\frac{5}{8}$ mile is about 1 km

Mass

1 ounce is about 28 grams

2.2 pounds is about 1 kilogram

Capacity

$1\frac{3}{4}$ pints is about 1 litre

1 gallon is about 4.5 litres

Calum's grandmother has given him this cake recipe.

All Calum's cookery equipment is modern and uses metric units.

He works out the size of the tin in centimetres.

$$1\,\text{inch} = 2.54\,\text{cm}$$
$$8\,\text{inches} \approx 8 \times 2.5\,\text{cm}$$
$$= 20\,\text{cm}$$

He works out how much flour he needs.

$$1\,\text{oz} \approx 30\,\text{g}$$
$$\text{so}\ 4\,\text{oz} \approx 4 \times 30\,\text{g} = 120\,\text{g}$$

1 ounce is about 28 grams and 28 is nearly 30

 *How much butter does he need?*

Next Calum works out how much milk he needs.

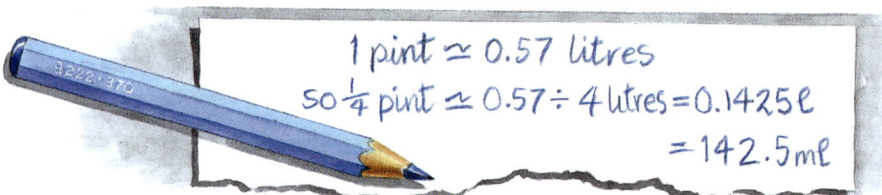

$$1\,\text{pint} \approx 0.57\,\text{litres}$$
$$\text{so}\ \tfrac{1}{4}\,\text{pint} \approx 0.57 \div 4\,\text{litres} = 0.1425\,\ell$$
$$= 142.5\,\text{m}\ell$$

 Many recipe books have the quantities listed in both Imperial and metric units. Usually the books advise you to use either Imperial or metric, not a mixture of both. Why?

1: Whole numbers and decimals

Exercise

1 Using the rough conversions on the opposite page, convert
   a) 60 gallons into litres
   b) 12 inches into centimetres
   c) 80 km into miles
   d) 44 lb into kg
   e) 30 litres into gallons
   f) 2 m into feet and inches
   g) 150 g into ounces
   h) 6 mm into inches.

2 Ewan drives out of Aberdeen and sees this distance sign (in miles).

He knows that 5 miles is 8 km.

How far in kilometres is
   a) Edinburgh?
   b) Glasgow?

| Edinburgh | 130 |
| Glasgow | 145 |

3 a) Gill weighs 9 stone 10 pounds. How many kilograms is this?
   b) Jeff weighs 11 stone 6 pounds. How many kilograms is this?

4 Convert these quantities as shown.

a)
Potatoes 10kg — Convert to pounds

b)
Mineral Water 5 litres — Convert to pints

c)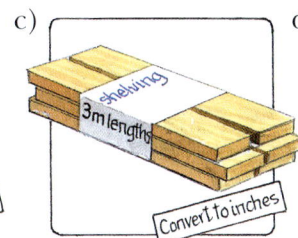
3m lengths shelving — Convert to inches

d)
SunFlower Margarine 200g — Convert to ounces

5 These are some of the specifications for Neil's car.

| Length | 5.7 m | Petrol tank capacity | 50 litres |
| Width | 2.2 m | Fuel consumption | 8 km per litre |

   a) Will Neil's car fit into a garage 18 feet long?
   b) What is the width of Neil's car in feet and inches?
   c) What is the petrol tank capacity to the nearest gallon?
   d) What is the fuel consumption in miles per gallon?

6 a)    b)

Write down each of these readings in grams, then convert it to pounds and ounces.

Some people say that certain cereal packets could be made smaller and so be less wasteful of the environment's resources.

Take 5 different cereal packets. Measure each one in cm and so work out the volume in cm^3. The weight of the cereal is marked on the packet in grams. Now work out the density in grams per cm^3 (weight ÷ volume). Which of your packets is the most tightly packed, and so the most environmentally friendly?

11

1: Whole numbers and decimals

Prime factorisation

 Which of these are prime numbers?

15, 17, 19, 21

If a number is not itself prime then it can be written as the product of primes.

e.g. 20 = 2 × 2 × 5   *Each of these is a prime*

and 84 = 2 × 2 × 3 × 7   *Each of these is a prime*

This is called **prime factorisation** (or **prime factor decomposition**).

Ian and Lin are both finding the prime factorisation of 20.

Ian writes this.   Lin writes this.

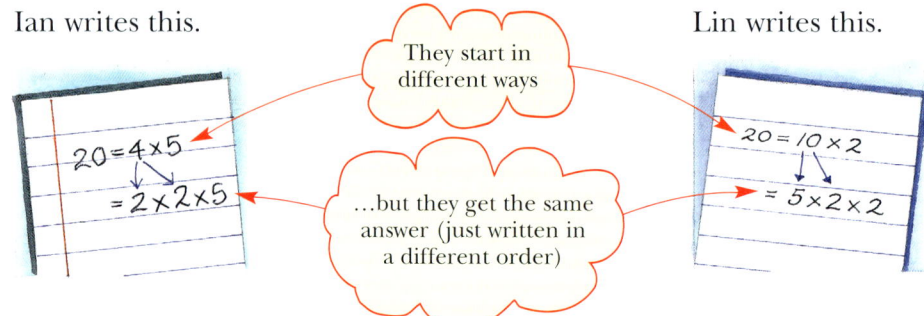

They start in different ways

...but they get the same answer (just written in a different order)

Remember that you must go on factorising until all the numbers are primes. Sometimes it may take several lines of working.

 What is the prime factorisation of 360?

Highest common factor (HCF)

You can find the HCF of 12 and 20 like this:

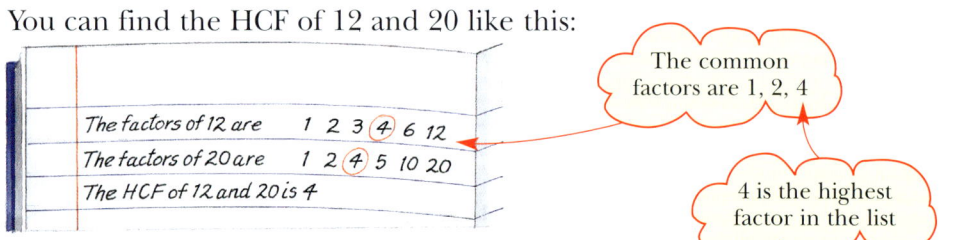

The common factors are 1, 2, 4

4 is the highest factor in the list

The HCF of 12 and 20 is 4.

Lowest common multiple (LCM)

You can find the LCM of 9 and 6 like this:

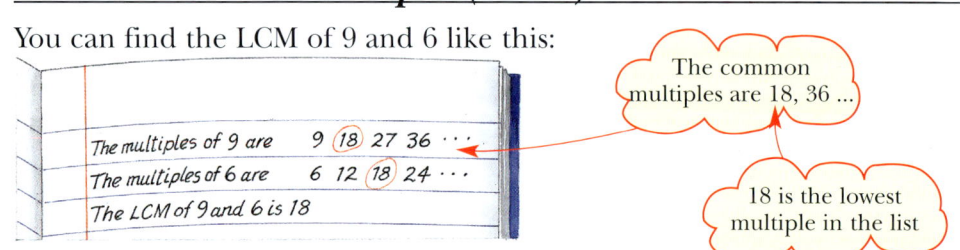

The common multiples are 18, 36 ...

18 is the lowest multiple in the list

1: Whole numbers and decimals

Exercise

1 Find the prime factorisation of each of these numbers.
   a) 14   b) 15   c) 28   d) 36
   e) 30   f) 27   g) 90   h) 126
   i) 150   j) 210   k) 539   l) 1540

2 Find the HCF of each of these.
   a) 6, 4   b) 6, 15   c) 18, 12   d) 12, 4
   e) 10, 25   f) 3, 8   g) 18, 45   h) 14, 10
   i) 21, 49   j) 22, 33   k) 63, 36   l) 56, 126
   m) 12, 24, 54 n) 25, 35, 75 o) 56, 24, 32   p) 60, 80, 100

3 Find the LCM of each of these.
   a) 10, 4   b) 5, 6   c) 4, 8   d) 12, 9
   e) 6, 10   f) 3, 7   g) 27, 18   h) 16, 8
   i) 14, 35   j) 8, 20   k) 20, 30   l) 45, 10
   m) 2, 3, 4 n) 5, 15, 2 o) 9, 12, 8 p) 6, 10, 15

4 a) Write down the LCM of 5 and 7.
   b) Write $\frac{2}{5}$ as a fraction with denominator 35.
   c) Write $\frac{3}{7}$ as a fraction with denominator 35.
   d) Which is larger, $\frac{2}{5}$ or $\frac{3}{7}$?

5 Look at these gear wheels.

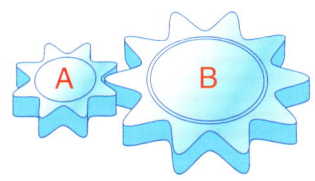

   a) A completes 10 turns. How many turns does B complete?
   b) What is the least number of turns that A can complete so that B also completes an exact number of turns?
   c) B completes 30 turns. On how many occasions will both A and B have been back in their starting position at the same time?

Roger has 4 hens.

Ingrid lays every second day. Ferdie lays every third day. Chookle lays every fourth day. Mumtie lays every fifth day. They all lay an egg on 1 January.

What is the date when they next all lay an egg?

On how many days in the year do they all lay an egg?

1: Whole numbers and decimals

Finishing off

Now that you have finished this chapter you should be able to

★ do calculations involving addition, subtraction, multiplication and division of whole numbers and decimals

★ do calculations using length, mass, capacity and money

★ convert between Imperial and metric units

★ work out simple squares, square roots, cubes and cube roots without a calculator

★ recognise primes and work out prime factorisations

★ recognise factors and find the HCF of two or more numbers

★ recognise multiples and find the LCM of two or more numbers

★ write very large and very small numbers using powers of 10.

Use the questions in the next exercise to check that you understand everything.

Mixed exercise

1 Work out

a) $2.4 \div 0.3$ b) 3.5×4.2
c) $9.3 \div 0.15$ d) 3.2×0.4

2 Find the value of

a) 3×10^4 b) $\sqrt{100}$
c) 7×10^{-2} d) $\sqrt[3]{125}$

3 Zach has £25 to spend on cassettes.

a) How many cassettes at £2.25 each can he buy?
b) How many cassettes at £1.75 each can he buy?
c) What combination of cassettes costs £25 exactly? Explain how you reached your answer.

4 Veena gets these quotes for her treasure hunt packs.

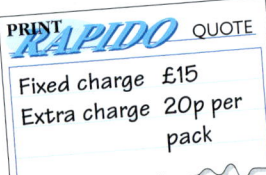
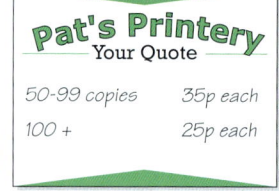

a) What is the lowest cost for 80 packs?
b) What is the lowest cost for 275 packs?
c) When does Print Rapido become cheaper than Pat's Printery?

1: Whole numbers and decimals

Mixed exercise

5 Convert
   a) 2.4 cm into mm
   b) 750 g into kg
   c) half a mile into yards
   d) 13.6 km into m
   e) 135 lb into stones and lb
   f) 1.5 litre into cl.

6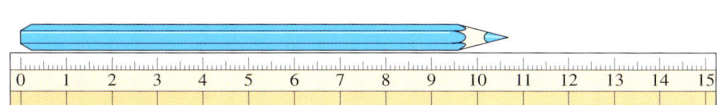

This is a 15 cm ruler. Write down the length of the pencil in
   a) centimetres   b) millimetres.

7 These numbers are in standard form. Write them out in full.
   a) the rest mass of an electron, 9.1×10^{-31} kg
   b) the population of Asia in 1900, 8.25×10^8
   c) the diameter of a human red blood cell, 7.5×10^{-6} m
   d) the escape velocity from Earth, 1.1×10^4 m s^{-1}

8 Write these numbers in standard form.
   a) the radius of the moon, 1 700 000 m
   b) the mass of an elephant, 20 000 kg
   c) the mass of a pygmy shrew, 0.0025 kg
   d) the length of a human chromosome, 0.000 005 m

9 Convert these Imperial measures into metric ones.
   a) 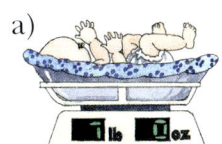   b) 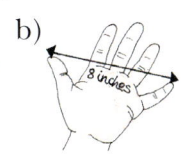   c)

10 Look at this list of numbers.
3, 6, 9, 27, 29, 36
Write down the ones that are
   a) factors of 27   b) multiples of 12
   c) primes   d) squares
   e) cubes.

11 Kelly goes to a disco every fourth Saturday, and ten-pin bowling every third Sunday. She does both during the weekend of 1 and 2 March.
   a) How many weeks pass before the two outings again occur in the same weekend?
   b) What will be the dates?
   c) How many more times before the year end will Kelly have the two outings in the same weekend?

12 Work out the prime factorisation of
   a) 18   b) 48   c) 100   d) 120

13 Write down the highest common factor of
   a) 15, 20   b) 16, 36
   c) 80, 30   d) 24, 36, 60

14 Write down the lowest common multiple of
   a) 4, 6   b) 24, 8
   c) 20, 50   d) 6, 10, 18

Identify six objects that are squares or contain squares. Measure their lengths using both metric and Imperial units. Present your results in a suitable way.

Two

Shapes and angles

Before you start this chapter you should

★ know that isosceles triangles have 2 equal sides and 2 equal angles

★ know that equilateral triangles have 3 equal sides and 3 equal angles

★ recognise different types of quadrilateral

★ know that angles round a point add up to 360° and angles on a straight line add up to 180°

★ be able to find pairs of equal angles where two lines cross

★ know that the angle sum in a triangle is 180° and in a quadrilateral is 360°.

Reminder

Use the questions in the next exercise to check that you still remember these topics.

- Where two lines intersect, opposite angles are equal.

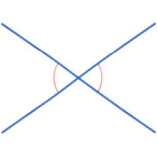

- Where a line intersects with two parallel lines, corresponding angles are equal.

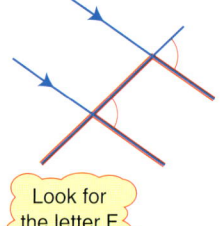

Look for the letter F

- Where a line intersects with two parallel lines, alternate angles are equal.

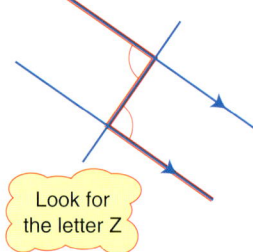

Look for the letter Z

Revision exercise

1 Find the angles marked with letters in these diagrams. The diagrams are not drawn accurately.

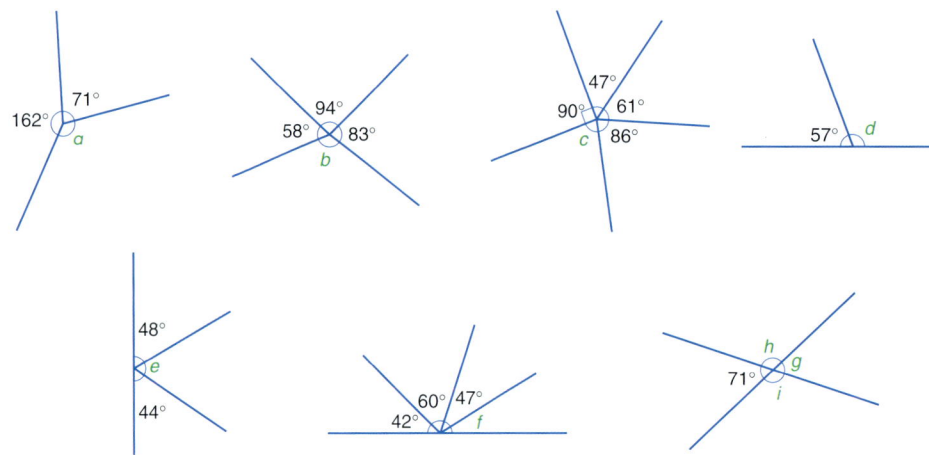

2: Shapes and angles

2 Find the angles marked with letters in these triangles.

a)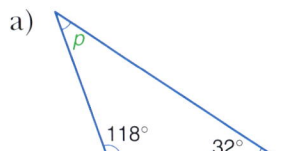
b)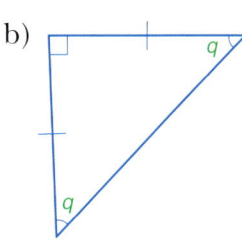
c)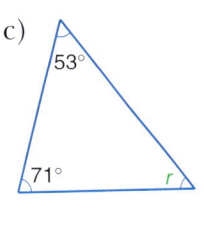
d)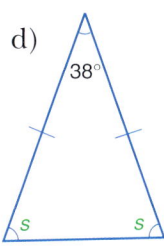

e)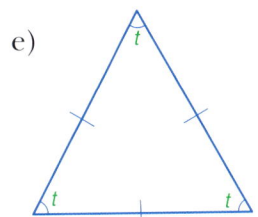
f)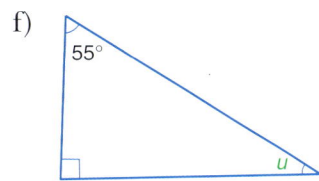
g)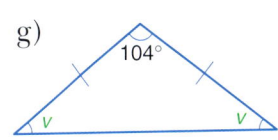

3 Describe each triangle above by two words, one from list A and one from list B.

　A equilateral, isosceles, scalene

　B right-angled, acute-angled, obtuse-angled

4 Find the angles marked with letters in these diagrams.

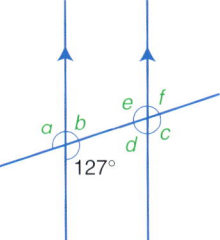 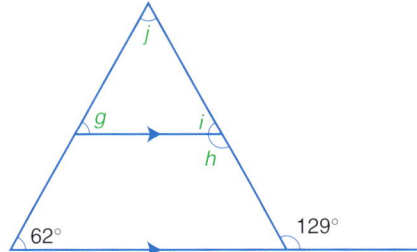

5 Look at this diagram.

Write down as many pairs as you can of

a) opposite angles
b) corresponding angles
c) alternate angles.

 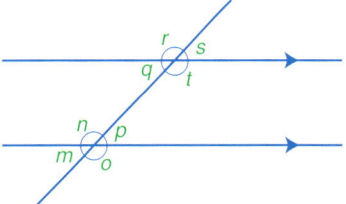

Here is a nonagon, made by joining nine equally spaced points on the arc of a circle.

Join each vertex to the centre of the circle and use the diagram to find the size of the interior angles of the nonagon.

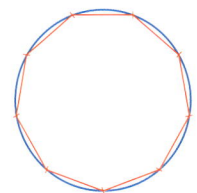

2: Shapes and angles

Quadrilaterals

Simon drew a triangle ABC and joined A to K, the mid-point of BC.
He then rotated the triangle about K.

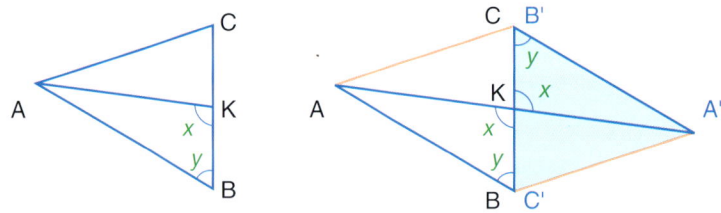

? Why are the angles marked y equal?
What are angles in the Z shape called?
Does this show that AB and A'B' are parallel?
Are the lines AB and A'B' equal?

? The lines AC and A'C' look parallel.
How can you be certain whether they are?
Are the lines AC and A'C' equal?

? Why are the angles marked x equal?
What are angles in this X shape called?
Is AKA' a straight line?
Is K the mid-point of AA'?

? What is the shape ABA'C called?

? What is the shape called if AB = AC?

? What is the shape called if CAB = 90°?

? What is the shape called if AB = AC and CAB = 90°?

? Suppose Simon had reflected the triangle in the line BC.
What would the shape be called then?
Would the diagonals be at right angles?
Would the diagonals bisect each other?
What would the shape be called if AB = AC?

2: Shapes and angles

1 Find the angles marked with letters in these quadrilaterals. Name each quadrilateral.

a) 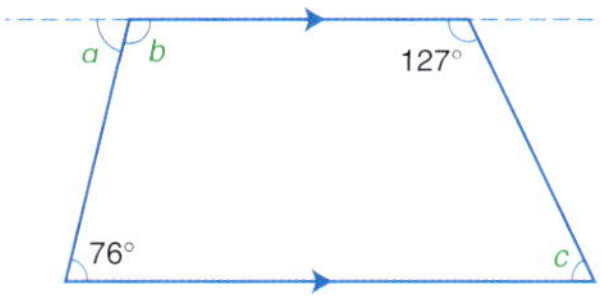 [127°, 76°, a, b, c]
b) 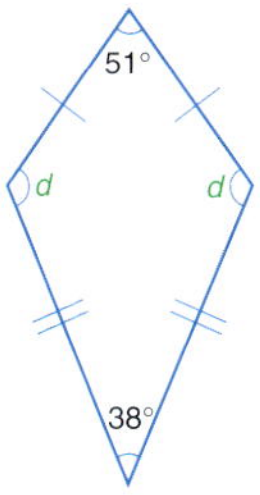 [51°, 38°, d, d]
c) 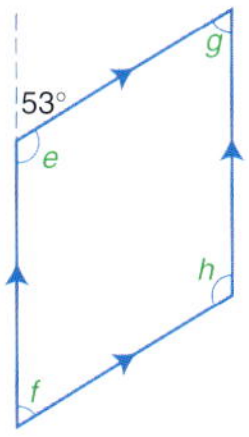 [53°, e, f, g, h]
d) 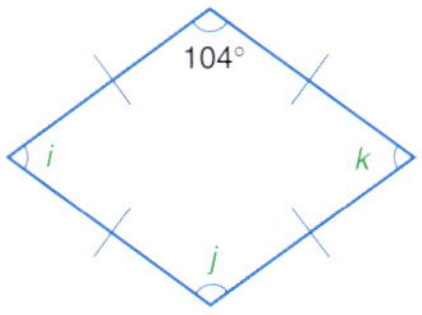 [104°, i, j, k]
e) 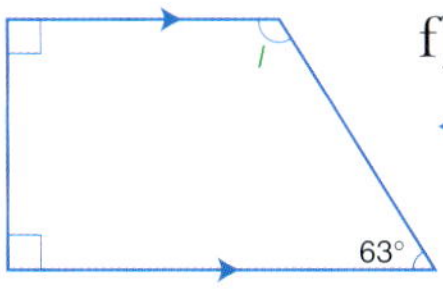 [63°, l]
f) [43°, 116°, m, m]

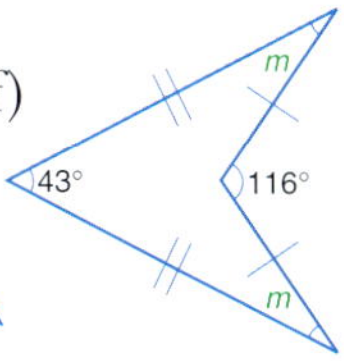

2 Find the angles marked with letters in these diagrams

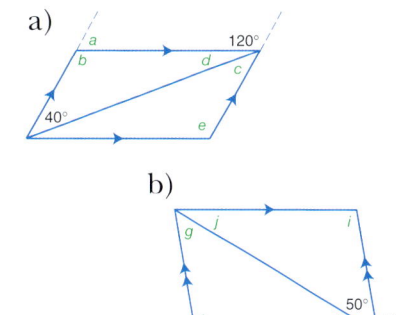

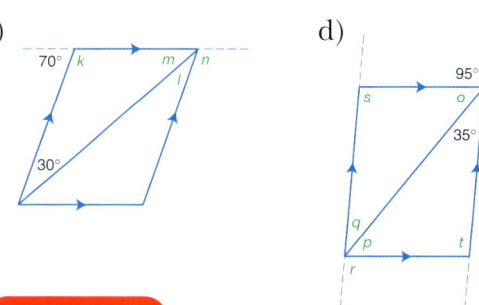

3 ABCD is a parallelogram. Name the special sort of parallelogram it is if

a) AC = BD
b) AC is perpendicular to BD
c) AC = BD and AC is perpendicular to BD.

4 Find the angles marked with letters in the quadrilaterals below.

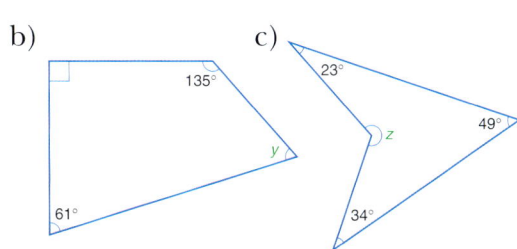

A parallelogram   B rhombus
C rectangle   D square
E trapezium   F kite

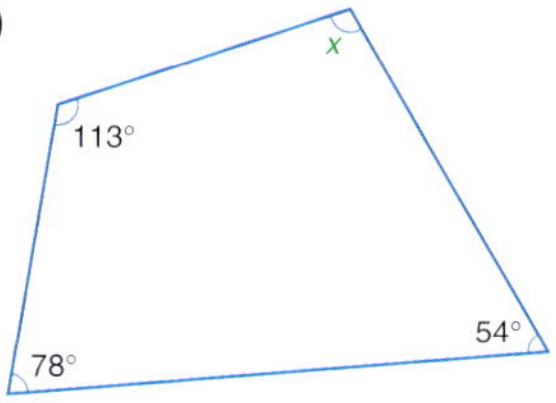

For which types of quadrilaterial is each of the following statements true?

a) Both pairs of opposite sides are equal.
b) All four sides are equal.
c) One (and only one) pair of sides is parallel.
d) Both pairs of opposite angles are equal.
e) One (and only one) pair of opposite angles is equal.
f) All angles are 90°
g) The diagonals are equal.
h) The diagonals are perpendicular.
i) The diagonals bisect each other.
j) The diagonals do not bisect each other.
k) The diagonals bisect the angles at the vertices.

2: Shapes and angles

Finishing off

Now that you have finished this chapter you should

★ be able to find pairs of equal angles where two lines cross and where a line intersects parallel lines

★ know what is meant by the words *quadrilateral, trapezium, parallelogram, rectangle, rhombus, square* and *kite*.

Use the questions in the next exercise to check that you understand everything.

Mixed exercise

1 Match these shapes with the descriptions below.

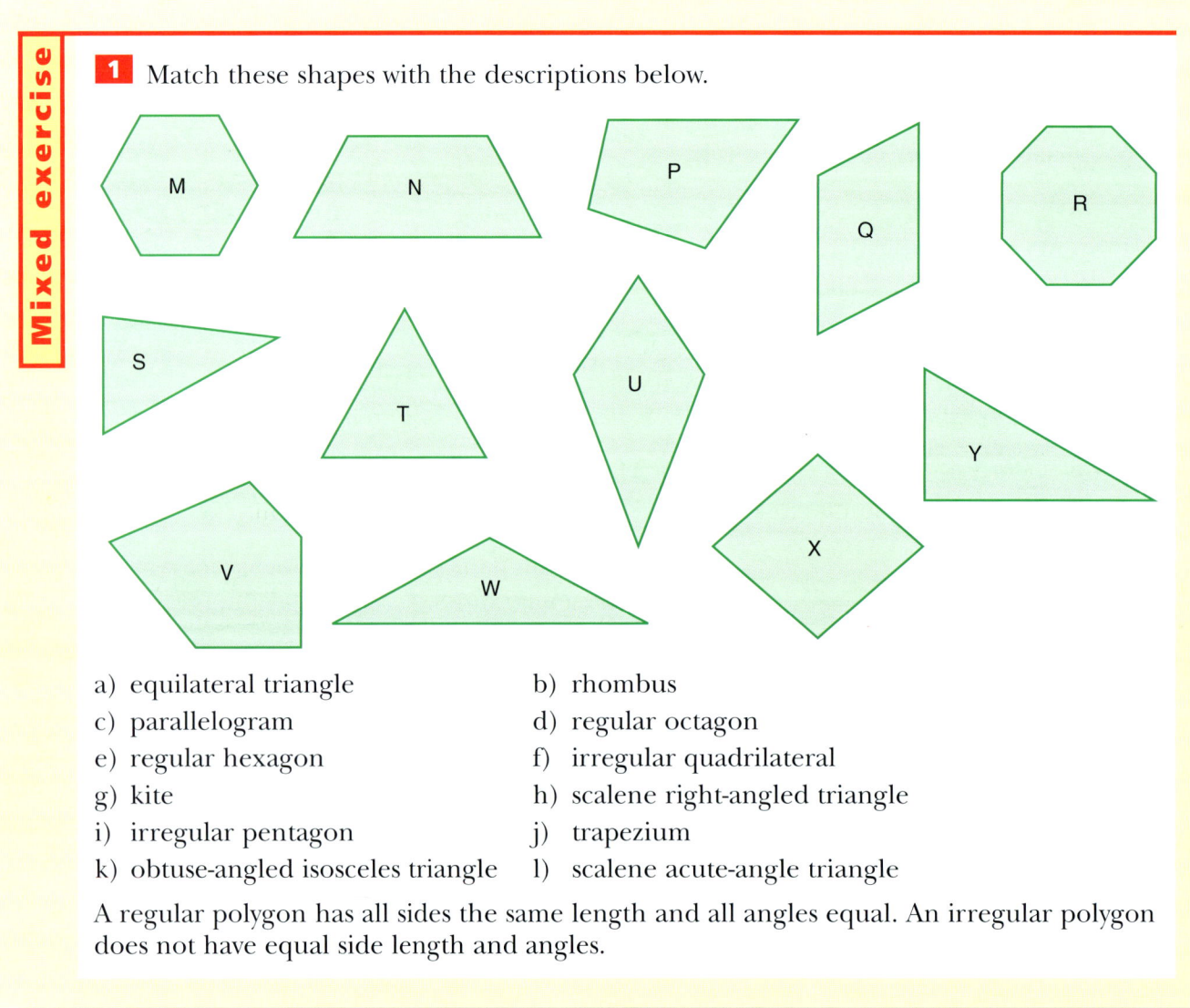

a) equilateral triangle
b) rhombus
c) parallelogram
d) regular octagon
e) regular hexagon
f) irregular quadrilateral
g) kite
h) scalene right-angled triangle
i) irregular pentagon
j) trapezium
k) obtuse-angled isosceles triangle
l) scalene acute-angle triangle

A regular polygon has all sides the same length and all angles equal. An irregular polygon does not have equal side length and angles.

2: Shapes and angles

2 Find the angles marked with letters in these diagrams.
(The diagrams are not drawn accurately.)

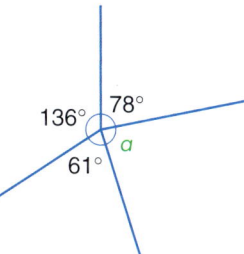 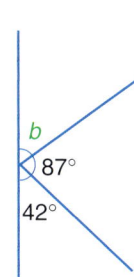 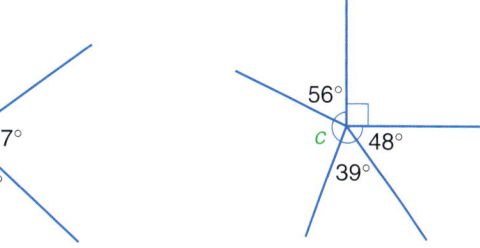

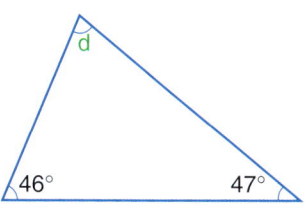 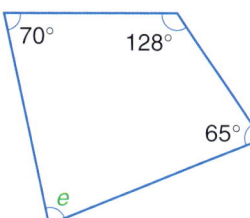 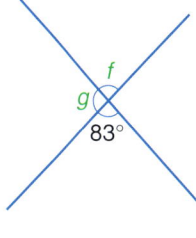

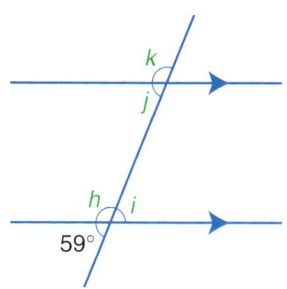 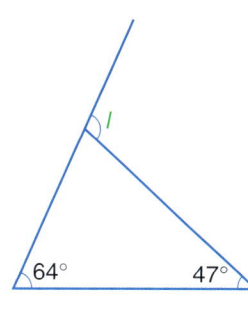 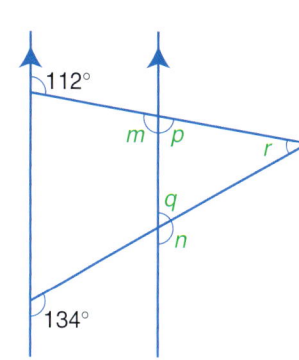

3 Find all the angles marked with letters in this diagram. (The dot is the centre of the circle.)

4 Find the size of the angles marked with letters in this diagram.

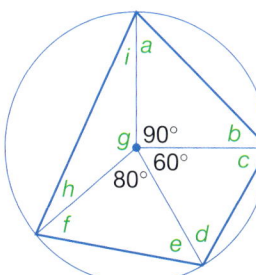 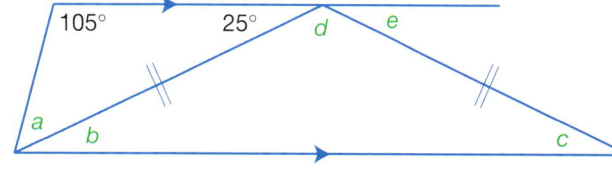

Mixed exercise

21

Three

Starting algebra

Writing things down

Jason has just woken up.

He has only a few minutes to get dressed, wash, eat breakfast and catch his bus.

 *Does it matter in what order he gets ready?*

When you are doing calculations, it is important to take the various steps in the right order.

Even calculators can get the wrong answer.

Most simple calculators will tell you that $3 + 4 \times 5$ is 35.

Scientific calculators will give 23.

 How do the two answers arise? Which one is right?

You can use the word BIDMAS to remind you of the right order for the steps in a calculation.

First — **B**rackets
Indices (or powers)
Divide
Multiply
Add
Last — **S**ubtract

Now you can be sure which answer is right.

$$3 + 4 \times 5$$

First multiply: $3 + 20$

Then add: 23

The right answer is 23.

Janet and Chris are taking their 2 children to the cinema.

 Which of these are the right ways to write down how much they pay?

A $2 \times 6 + 3$ B $6 + 3 \times 2$

C $2 \times 6 + 2 \times 3$ D $2 \times (6 + 3)$

 *How would you describe what brackets do?*

22

3: Starting algebra

1 Mr and Mrs Jones take their 2 grandchildren on a train to Blackpool.

Their tickets cost £15 each and the children's are £8 each.

Which of these are the right ways to write down how much they pay?

A $2 \times 15 + 8$
B $2 \times 15 + 2 \times 8$
C $15 + 8 \times 2$
D $2 \times (15 + 8)$

2 Mr and Mrs Jones are senior citizens. They want to take their 2 grandchildren to the cinema.

Write down two ways of working out how much it will cost them.

Use the prices on the opposite page.

In questions 3 to 5, work out the answer. Write each step underneath the previous one, as on the opposite page, so that you can see where the numbers come from.

3 a) $5 + 2 \times 3$ b) $2 \times 3 + 5$ c) $8 + 12 \div 4$
d) $22 - 4 \times 5$ e) $11 + 33 \times 2$ f) $10 - 9 \div 3$
g) $4 \times 5 - 18$ h) $4 \times 5 + 12 \div 3$

4 a) $(5 + 2) \times 3$ b) $2 \times (3 + 5)$ c) $(8 + 12) \div 4$
d) $(22 - 4) \times 5$ e) $5(12 - 7)$ f) $2 + 5(12 - 7)$
g) $13 - (7 - 2)$ h) $12 \div (6 \div 3)$

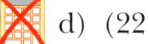

5 a) $(5 + 3)^2$ b) $5 + 3^2$ c) $(4 \times 5)^2$
d) 4×5^2 e) $7 + (2 \times 4)^2$ f) $7 + 2 \times 4^2$

*Remember to work out **B**rackets before **I**ndices*

6 Use a scientific calculator to check your answers to questions 3, 4 and 5. Put in the multiplication sign in parts 4e) and 4f), but otherwise enter each one exactly as it is written, including the brackets.

Make a list of all the actions required to make a cup of tea or coffee. Organise your list to show which actions must be done before others.

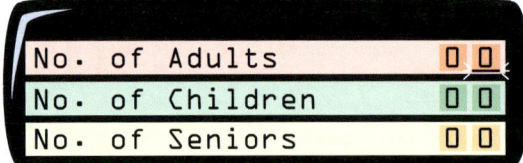

3: Starting algebra

The language of algebra

The box office for the cinema has a computer which works out the cost of tickets.

This is what it looks like when it is ready to use.

When the numbers of adults, children and senior citizens in a party have been keyed in, the computer works out

6 × number of adults + 3 × number of children + 2 × number of senior citizens

In algebra, you could write this as

$6a + 3c + 2s$

s stands for the number of senior citizens

a stands for the number of adults

c stands for the number of children

*a, c and s, are **variables**. Their values change (or vary) with each new group*

Notice that $6 \times a$ is written as $6a$. We can leave out the × sign in algebra.

 Why can't you leave out the × sign in arithmetic?

The **expression** $6a + 3c + 2s$ has 3 **terms**, $6a$, $3c$, and $2s$.

These are **unlike terms** because they involve different letters, a, c and s.

The terms in the expression $3x + 2x + 4x - x$ are **like terms** because they all have the same letter x.

x is just 1x

3 + 2 + 4 − 1 = 8

The expression $3x + 2x + 4x - x$ can be simplified to $8x$.

The expression $6a + 3c + 2s$ cannot be simplified any more. It is in its simplest form.

When 2 adults, 3 children and 1 senior citizen go to the cinema, $a = 2$, $c = 3$ and $s = 1$.

Substituting these into the expression $6a + 3c + 2s$ gives

$6 \times 2 + 3 \times 3 + 2 \times 1$
$= 12 + 9 + 2 = 23$

Remember BIDMAS, multiply before adding

It costs the family £23 to go to the cinema.

 Describe a family for which $a = 0$, $c = 4$ and $s = 2$.

3: Starting algebra

Exercise

1 Use the expression $6a + 3c + 2s$ to work out the cost of cinema tickets for

a) 3 adults, 4 children and 1 senior citizen;

b) a school party of 40 children and 3 teachers.

2 The following year the prices for the cinema are increased to £8 for an adult, £5 for a child and £4 for a senior citizen. Write down a new expression for working out the total bill for a group of people.

3 Copy each expression and write it as briefly as possible.

State how many terms there are in each simplified expression.

a) $5 \times a + 6 \times b$
b) $6 \times q - 3 \times p$
c) $4 \times w + 1 \times y + 7 \times z$
d) $1 \times r - 7 \times s + 1.5 \times t$
e) $4 \times x + 5 \times y + 1 \times z$
f) $2 \times k + 3 \times 4 \times n$
g) $1 \times 5 \times b$
h) $e \times 12 - 6 \times f$
i) $3 \times n - m \times 7 + p \times 4 + q$
j) $f \times 1 + 4 \times d - c \times 19$

Write this as 12e. The numbers come first in each term.

4 Copy each of these and write it more simply.

a) $6a + 7a$
b) $8c - 3c$
c) $2x + 3x + 4x$
d) $5y + 2y - 3y$
e) $12x - 5x - 3x$
f) $4d - d + 6d - 8d$

5 For each of these, put the like terms into separate lists, then add the terms in each list.

a) $x, 5x, 7, -5, 9x$
b) $a, 2b, 7a, 4b, 6b, -a$

6 You can write some expressions more simply by collecting together the like terms. For example,

$2a + 3b + a - b$
$= 2a + a + 3b - b$
$= 3a + 2b$

Copy each expression and collect together the like terms.

Write each expression as simply as possible.

a) $4a + 2a$
b) $4a + b + 2a$
c) $2x + 8y + 5x$
d) $2x + 2y + 5x + 2y$
e) $5n + 8n + 6 + 2n$
f) $7k + 5 + 4k + 2k + 5$
g) $y + 5x + 4y + 6x - 3y$
h) $5a + 4c + 9a - 3a - 2c - 3a$

A charity box contains a number of coins. Write down expressions for the number of coins, their total value and their weight.
(Hint: let a be the number of 1p coins, b be the number of 2p coins, c be the number of 5p coins, etc.)

3: Starting algebra

Substituting into a formula

Tom is looking up how long it takes to roast a goose. His cookery book says *45 minutes per kg plus 20 minutes*.

Tom's goose weighs 2.7 kg. For how long should he cook it?

Tom works it out like this:

> cooking time (minutes) is
> $2.7 \times 45 + 20$
> $= 121.5 + 20$
> $= 141.5$
> That's 2 hours $21\frac{1}{2}$ minutes

You can use algebra to write the formula. Calling the weight of the goose w kg and the cooking time t minutes, the formula would be

$t = 45w + 20$

You just substitute for the weight of the goose.

Look carefully at this formula.

Where does the '45w' come from?

Where does the '+ 20' come from?

Example

Tom has another goose of weight 3.2kg.

Use the formula to work out the cooking time for this one.

Solution

Substituting $w = 3.2$ into the formula gives

$t = 3.2 \times 45 + 20$

$= 144 + 20$

$= 164$

The cooking time is 2 hours 44 minutes.

3: Starting algebra

Exercise

1 Calculate s in each of these formulae by substituting the numbers given.
   a) $s = 4u + 80$ (i) $u = 2$ (ii) $u = 5$ (iii) $u = 6.5$
   b) $s = 4u + 8a$ (i) $u = 2, a = 5$ (ii) $u = -6.5, a = 10$

2 The area, A, of a trapezium is given by
$$A = \frac{1}{2} h (a + b)$$
Find the value of A when
   a) $a = 6, b = 4, h = 2$
   b) $a = 16, b = 4, h = 5$
   c) $a = 8, b = 8, h = 8$

3 The volume, v m³, of gravel required for a rectangle of dimensions w metres by l metres is given by the formula
$$v = 0.05 lw$$
Find the volume required for
   a) a path 1 m wide and 20 m long
   b) a driveway 2.5 m wide and 12.5 m long.

4 When a stone is dropped down a well, you can use the formula
$$d = 5t^2$$
to estimate its depth d in metres, where t is the time in seconds it takes to reach the bottom. Find d when $t = 4$.

5 Clare is making curtains. The length of fabric she needs for a window d metres deep is l metres, where
$$l = n(d + 0.5).$$
The number n depends on the width of the window.

How many metres of fabric must Clare buy when $n = 2$ and $d = 2.5$?

6 When a ball is thrown straight up in the air at 5 metres per second, its *upward* speed v metres per second after t seconds is given by the formula
$$v = 5 - 10t$$
Find v when t is:
   a) 0   b) 0.2   c) 0.5   d) 1

Write down what is happening to the ball in each case.

You would use the formula for cooking a goose that is given on the opposite page, $t = 45w + 20$, if you had metric scales.

The rule usually used for a goose weighed in pounds is 'Twenty minutes plus twenty minutes per pound'. Write this as a formula.

Draw a graph with two lines on it, one for each formula. Cooking time should be on the vertical axis and the weight of the goose on the horizontal axis. The horizontal axis should be marked in both kilograms and pounds.

27

3: Starting algebra

Using brackets

Gus and Dougie are buying crisps for a party. They decide to get 3 of these special-offer bags.

How many packets of crisps will they have altogether?

Gus and Dougie have different ways of working it out.

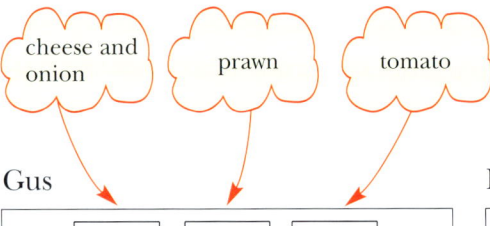

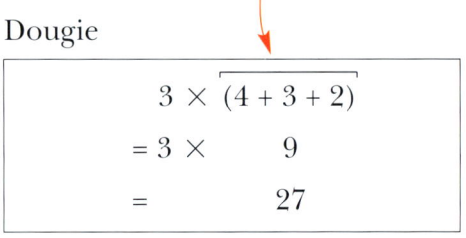

You can see that they both get the same answer:

$$3 \times 4 + 3 \times 3 + 3 \times 2 = 3 \times (4 + 3 + 2)$$

If they buy 5 special offer bags, the number of packets will be

$$5 \times 4 + 5 \times 3 + 5 \times 2 = 5 \times (4 + 3 + 2)$$

You can write the same for any number of bags, say n:

$$n \times 4 + n \times 3 + n \times 2 = n \times (4 + 3 + 2)$$

or, more neatly
$4n + 3n + 2n = n(4 + 3 + 2)$

 *Check that this works for $n = 6$.*

A different special-offer bag contains c packets of cheese and onion, p packets of prawn and t packets of tomato.

 *What does $c + p + t$ represent?*

Describe the meaning of $6(c + p + t)$.

Describe what each term represents in $6c + 6p + 6t$.

 *The total number of packets is $n(c + p + t)$.*

How many packets are there when $n = 6$, $c = 5$, $p = 4$ and $t = 3$?

 *Write an expression equivalent to $4(x + 2y - 3z)$.*

*What you have done is called **expanding the brackets**.*

3: Starting algebra

Exercise

1 Simplify these expressions.

a) $x(5 + 2)$ b) $(6 + 4 + 3)y$
c) $n(22 + 77 + 1)$ d) $17x + 3x + 10x$

Remember to write the number first in the answer

2 Copy each of these and expand the brackets.

a) $3(a + b)$ b) $5(c + d + e)$ c) $10(x + 2)$
d) $2(3 + y + z)$ e) $7(f + 3 + g)$ f) $2(x - 5)$
g) $4(x - y)$ h) $(p + q - r) \times 2$ i) $(c + d + 1) \times 8$

3 a) What is the value of $2m + 3$ when $m = 5$?

b) What is the value of $4(2m + 3)$ when $m = 5$?

c) Expand $4(2m + 3)$.

d) Find the value of your expanded expression when $m = 5$. Check that your answer is the same as the one for b).

4 a) What is the value of $(3x - 2y)$ when $x = 4$ and $y = 2$?

b) What is the value of $5(3x - 2y)$ when $x = 4$ and $y = 2$?

c) Expand $5(3x - 2y)$.

d) Find the value of your expanded expression when $x = 4$ and $y = 2$. Check that your answer is the same as the one for b).

5 Simplify each of these by expanding the bracket then adding the like terms. Check each one by substituting $x = 4$ and $y = 3$ in both the question and your answer; both should work out to be the same.

a) $2(x + y) + 3y$ b) $3(x + 5) - 7$ c) $10(x + 3) + 2x$
d) $4(5 + x) + 8$ e) $2(y + 2x) + x$ f) $6(x + y) + 2x$
g) $4(x + 4y) + 2x$ h) $7(x + y + 7) + y + 7$

6 Write these using brackets

a) $3x + 3y + 3z$
b) $3x + 6y + 9z$
c) $6a - 3b$
d) $8a - 4b + 16c$
e) $36p - 12q + 18r$
f) $100u - 25v - 75w$

Write down the key strokes you need to enter $3 \times 5 + 3 \times 6$ into your calculator
a) as it is written and
b) using brackets.
How many key strokes are involved in each?
What about $3 \times 5 + 3 \times 6 + 3 \times 7$?
How many key strokes do you save by using brackets in a long calculation like this?

3: Starting algebra

Adding and subtracting with negative numbers

George works in the Hillside Hotel.

He takes a bottle of champagne in the lift from the cellar (floor –5) to the honeymoon suite (floor 4).

The diagram shows George's trip.

His journey can be written as

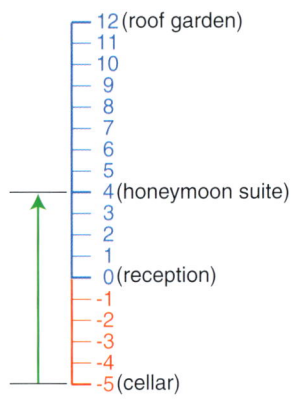

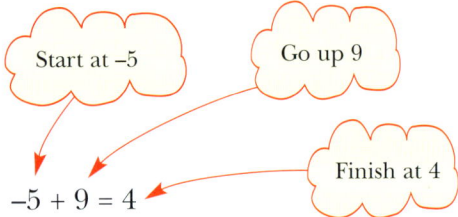

−5 + 9 = 4

The honeymoon couple are in the roof garden at floor 12. They take the lift down to their suite. The lift stops on the way to let someone else get in. Here is an expression for the honeymoon couple's trip down from the rooftop garden:

12 − 5 − 3 = 4

 *Draw a diagram to show their trip.*

At what floor did the lift stop on the way?

You could show these trips on a horizontal number line to save space.

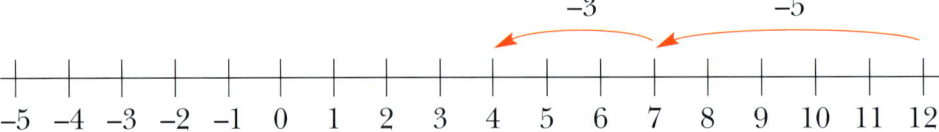

 *What does 12h − 5h − 3h come to?*

What does − 3h − 5h + 12h come to?

 *If h is the distance from one floor to the next, what does 4h represent?*

 *How do you key in −5 on your calculator?*

Work out −5 + 4, −3 + 12 and −4 − 6 on your calculator.

Check your answers on a number line.

30

3: Starting algebra

Exercise

1 Draw a number line to illustrate each of these, and write down the answer to each.

a) 6 − 4 b) −4 + 6 c) −6 + 4
d) 4 − 6 e) 2 + 3 − 5 f) 2 − 4 − 1
g) −11 + 3 + 5 h) −18 − 21 i) −55 + 61

2 There are many ways of making 7 using the numbers 12, 4 and 1.

For example, −4 − 1 + 12 = 7.

Write down 3 other ways of making 7 using these numbers.

Check each way using a calculator and a number line.

3 Write these more simply.

a) $6x - 4x$ b) $4x - 6x$ c) $-4x - 6x$
d) $7y - 3y$ e) $-3y + 7y$ f) $3y - 7y$
g) $2x + 3x - 4x - x$ h) $2y - 4y + 3y$ i) $-7y - 3y - 15y$

4 Substitute $x = 2$ and $y = 3$ in the expressions in question 3, and in each of your answers, and check that you get the same results.

5 Write these as simply as possible.
(Remember that you can only add and subtract *like terms*.)

a) $2 + 3 - 5p$ b) $6q - 3q + 4 - 3$ c) $p + 5 - 11$
d) $6 - 4p - 3p$ e) $7 + 5q - 4 + 6q$ f) $8p + 5 - 9p - 2$
g) $q + 3p - 4q + p$ h) $-46p - 18q + 46p$ i) $-3 + q + 3 - q$

6 Substitute $p = 1$ and $q = 2$ in the expressions in question 5, and in each of your answers, and check that you get the same results.

A lift sometimes goes past the floor you want, or stops at another floor on the way.

Write 4 different ways for George at the Hillside Hotel (see opposite page) to get to the kitchen on floor −3 after delivering the champagne on floor 4.

7 Write a set of instructions to enable a friend to use your calculator for adding and subtracting negative numbers.

3: Starting algebra

Finishing off

Now that you have finished this chapter you should

★ know that
  $1 \times n = n$, $2 \times n = 2n$,
  $3 \times 2n = 6n$

★ understand the words *expression* and *term*

★ be able to simplify an expression, doing the steps in the right order

★ be able to substitute into a formula

★ be able to work with brackets and expand them

★ be able to add and subtract like terms

★ be able to use a number line for negative numbers.

Use the questions in the next exercise to check that you understand everything.

Mixed exercise

1 Work out the value of each of these.

a) $4 + 6 \times 2$
b) $(4 + 6) \times 2$
c) $12 - 6 + 1$
d) $12 - (6 + 1)$
e) 3×2^2
f) $(3 \times 2)^2$
g) $2n^2$ when $n = 5$
h) $(2n)^2$ when $n = 5$
i) $12 + 3 \times 4 - 2 \times 8 + 6 \div 2$
j) $8 - 2 \times 3 + 4 \div 2 - 1$
k) $2(x + 6)$ when $x = 3$
l) $2x + 12$ when $x = 3$
m) $3(9 - 2x)$ when $x = 2$
n) $27 - 6x$ when $x = 2$

2 Copy the following expressions and write them as simply as possible.

State how many terms there are in each.

a) $3 \times a + 6 \times b$
b) $12 - 6 \times x$
c) $2 \times y - 5 \times z + 4$
d) $3 \times 5 \times c$
e) $1 \times p + 1 \times q + 1 \times r$
f) $99 \times f - g \times 12$
g) $3 \times x - 6 \times y + 5 \times 3$
h) $n \times 7 - 3 \times m$
i) $3 \times q + 7 \times p$
j) $5 \times r + 6 \times s - 2 \times r - 5 \times s$

3: Starting algebra

Mixed exercise

3 Copy each expression and collect together the like terms.
Write each expression as simply as possible.
- a) $4a + 2a$
- b) $4a + 1 + 2a$
- c) $2x - 8 + 5x$
- d) $2x + 8 - 5x$
- e) $5n + 8n - 6 + 2n$
- f) $7k - 2 - 4k - 2k + 5$
- g) $p + 4x + 7y - 6x + 13p$
- h) $9a + 5c - 3d - 2c - 3a$
- i) $101d - 52w + 97 - 7d + w$
- j) $-4a - 16 - 13a + 7$

4 Alf has a decorating business and he can claim back the VAT that he spends on materials. When he spends £P the VAT is given by the formula
$$V = \frac{7}{47} \times P$$
Find the amount of VAT he can claim for paint costing £23.50.

5 Work out the time Mac should cook a turkey weighing 6.3 kg by:
- a) using the rule *15 minutes per 450 g plus 15 minutes*;
- b) using the formula $T = 33W + 15$ where T is the time in minutes for a turkey weighing W kg. Is your answer close enough to a)?

6 Copy and expand these expressions.
- a) $2(n - 5)$
- b) $2(7 + 3x)$
- c) $(3a + 6)\ 4$
- d) $3(4y + 2z)$
- e) $5(6 - a)$
- f) $(a + b + c) \times 3$

7 Copy and simplify these expressions.
- a) $2(x + 8) + 1$
- b) $12 + (5 + y)$
- c) $3 + 4(2n - 1)$
- d) $(6d + 5) + d$
- e) $7 + 6(4a - 2)$
- f) $3x + 9\ (5 - x)$
- g) $2(x + 3y) + 3(x + y)$
- h) $2p + 3(p + q)$
- i) $12(2p - 5q) - 6(2p + p)$
- j) $3(2s + 4r + 11) - 6s - 2r - 33$

8 a) Substitute $a = 4$, $b = 5$ and $c = 6$ into
   (i) $3a + 12b + 6c$
   (ii) $3(a + 4b + 2c)$
b) Why are your answers the same?
c) Check that you get the same answers when you substitute $a = 2$, $b = 1$ and $c = -3$ into the two expressions.

Investigation

You have some scales and a 1 g, a 3 g and a 9 g weight.

Explain how you can use them to weigh objects of 1 g, 2 g, ..., 13 g.

You are allowed a fourth weight. What do you choose?

Four

Fractions and percentages

> **Before you start this chapter you should be able to**
>
> ★ find equivalent fractions
> ★ write a fraction in its simplest form
> ★ find the reciprocal of a number
> ★ convert between improper fractions and mixed numbers
> ★ add and subtract fractions
> ★ multiply a whole number by a fraction.

Reminder

- The top line of a fraction is called the **numerator**.
- The bottom line of a fraction is called the **denominator**.
- There is more than one way of writing the same fraction. For example,

    $$\frac{1}{4} = \frac{2}{8} = \frac{3}{12} = \frac{4}{16} = \ldots\ldots$$

    These are called **equivalent fractions**.

- $\frac{1}{4}$ is a fraction in its **simplest form** or **lowest terms**.

- The **reciprocal** of 4 is $\frac{1}{4}$; the **reciprocal** of $\frac{1}{4}$ is 4.

- $\frac{21}{8}$ is an **improper fraction**, or top heavy fraction.

- $2\frac{5}{8}$ is a **mixed number**.

Revision exercise

1 Find the missing number in each of these.

a) $\frac{1}{2} = \frac{?}{6}$
b) $\frac{3}{4} = \frac{?}{8}$
c) $\frac{40}{50} = \frac{4}{?}$
d) $\frac{7}{8} = \frac{?}{16}$
e) $\frac{1}{4} = \frac{25}{?}$
f) $\frac{7}{10} = \frac{?}{100}$
g) $\frac{65}{100} = \frac{?}{20}$
h) $\frac{3}{4} = \frac{?}{52}$

2 Write each fraction in its simplest form.

a) $\frac{12}{16}$
b) $\frac{5}{20}$
c) $\frac{6}{15}$
d) $\frac{16}{24}$
e) $\frac{36}{60}$
f) $\frac{35}{56}$

4: Fractions and percentages

Revision exercise

3 a) Write down the reciprocal of (i) 5 (ii) $\frac{1}{5}$. What is $5 \times \frac{1}{5}$?

b) Write down the reciprocal of (i) $\frac{1}{10}$ (ii) 10. What is $\frac{1}{10} \times 10$?

c) Write down the reciprocal of 1.

d) Why does 0 not have a reciprocal?

4 a) Find the missing number in $\frac{2}{3} = \frac{?}{12}$.

b) Find the missing number in $\frac{3}{4} = \frac{?}{12}$.

c) Which is larger, $\frac{2}{3}$ or $\frac{3}{4}$?

d) What is the difference between $\frac{2}{3}$ and $\frac{3}{4}$?

5 Work out

a) $\frac{1}{2} \times 220$
b) $\frac{1}{3} \times 45$
c) $\frac{2}{3} \times 90$
d) $\frac{5}{7} \times 550$

6 Change these improper fractions to mixed numbers.

a) $\frac{9}{2}$
b) $\frac{13}{8}$
c) $\frac{12}{5}$
d) $\frac{11}{3}$
e) $\frac{15}{4}$
f) $\frac{13}{6}$

7 Change these mixed numbers into improper fractions.

a) $3\frac{1}{2}$
b) $4\frac{3}{8}$
c) $1\frac{7}{16}$
d) $2\frac{3}{4}$
e) $5\frac{1}{3}$
f) $3\frac{11}{16}$

8 In each of these, work out the answer and write it as a mixed number in its simplest form.

a) $3\frac{3}{8} + \frac{7}{8}$
b) $1\frac{1}{4} + 3\frac{5}{8}$
c) $3\frac{7}{8} - 1\frac{5}{16}$
d) $4\frac{1}{2} - 2\frac{7}{8}$
e) $4 - \frac{5}{8}$
f) $2\frac{3}{4} + 1\frac{5}{8}$
g) $5\frac{1}{4} - 3\frac{5}{8}$
h) $3 - \frac{11}{16}$
i) $1\frac{3}{4} + 2\frac{7}{16}$

9 Arrange these numbers in order of size, smallest first.

a) $4 \quad \frac{11}{3} \quad \frac{17}{4} \quad 4\frac{3}{16}$

b) $\frac{11}{4} \quad 3 \quad 2\frac{13}{16} \quad \frac{23}{8}$

Investigation

Write the fractions $\frac{1}{12}, \frac{2}{12}, \frac{3}{12}, ..., \frac{11}{12}$ in their simplest form.

What happens when you try to do this for elevenths?

What about ninths, tenths and thirteenths?

Predict what would happen if you tried to simplify the set of fractions with denominator 29 (i.e. different numbers of twenty-ninths.) Explain your answer.

4: Fractions and percentages

Using fractions

Victoria is organising a reception for 90 adults during the interval of a play. She needs to order the drinks for it. She is offering fruit juice, white wine or red wine.

Victoria estimates that $\frac{1}{3}$ of the people will drink fruit juice. Here is her calculation.

Notice that she writes × for 'of'

Juice $\frac{1}{3} \times 90 = 30$ $3\overline{)90} = 30$

She divides 90 by 3 to get 30

? She allows half a carton of juice per person. How many cartons is this? Who will drink fruit juice? Is $\frac{1}{3}$ a sensible estimate?

Victoria works out that 60 will drink wine.

Victoria knows from past experience that about $\frac{2}{3}$ of the people having wine will choose white and the rest red.

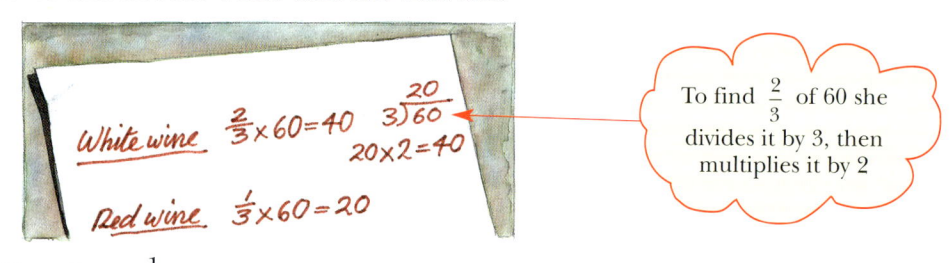

White wine $\frac{2}{3} \times 60 = 40$ $3\overline{)60} = 20$ $20 \times 2 = 40$

Red wine $\frac{1}{3} \times 60 = 20$

To find $\frac{2}{3}$ of 60 she divides it by 3, then multiplies it by 2

She allows $\frac{1}{4}$ of a bottle of wine per person. How many bottles of white wine and red wine is this?

Out of the 90 people, Victoria expects 20 to have red wine. This is $\frac{20}{90}$ or $\frac{2}{9}$ of the people.

On this pie chart it is $\frac{2}{9} \times 360° = 80°$.

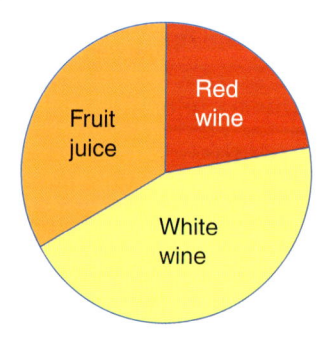

? What are the angles for the fruit juice and white wine sectors on this pie chart?

36

4: Fractions and percentages

1 This map shows the distances, in miles, between six stations.

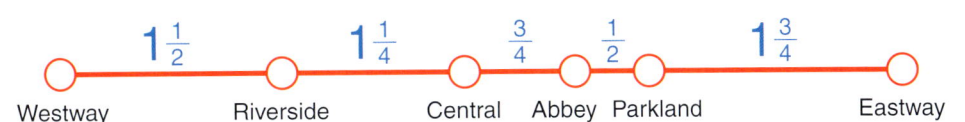

Westway —$1\frac{1}{2}$— Riverside —$1\frac{1}{4}$— Central —$\frac{3}{4}$— Abbey —$\frac{1}{2}$— Parkland —$1\frac{3}{4}$— Eastway

What is the distance from

a) Central to Parkland?
b) Westway to Central?
c) Riverside to Abbey?
d) Westway to Eastway?

2 Anna is a cook. She cuts each chicken pie into 16 portions.

a) How many people can she serve with $2\frac{1}{2}$ pies?

b) How many pies would she need to serve 60 people?

3 Nurse Ryan runs an asthma clinic. He allows $\frac{1}{4}$ hour for each appointment.

a) How many appointments can he fit in between 1330 and 1700?

b) Only the first 9 slots are taken. At what time does the last patient leave?

4 Elizabeth buys this jacket.

a) How much does she save by buying it in the sale?

b) Another shop prices the same jacket at £65, and reduces the price by a quarter in the sale. Is this a better deal?

SALE $\frac{1}{3}$ off all marked prices!
Jacket £69·99

5 A magazine editor has a policy that three eighths of the pages carry colour photographs and three quarters carry advertisements.

The November magazine has 128 pages.

a) How many pages carry colour photographs?
b) How many pages do *not* carry advertisements?

For the December special issue the number of pages is increased by a quarter.

c) How many pages are there?
d) How many pages carry advertisements?

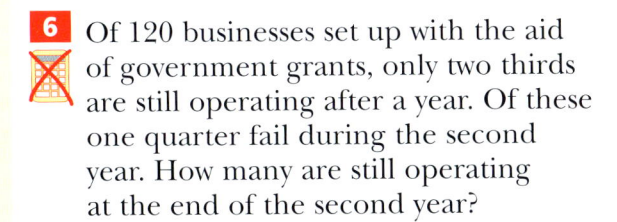

6 Of 120 businesses set up with the aid of government grants, only two thirds are still operating after a year. Of these one quarter fail during the second year. How many are still operating at the end of the second year?

Find a pie chart in a newspaper, magazine or brochure.
Work out what fraction of the total each sector represents.

4: Fractions and percentages

From fractions to percentages

A new treatment for asthma is tested on 50 sufferers. Of these, 35 find that it is better than their old treatment.

You can write 35 out of 50 as a fraction, $\frac{35}{50}$. Its simplest form is $\frac{7}{10}$. However, you often see results like this given as percentages.

There are two ways to change a fraction into a percentage. You need to be able to use both of them. Sometimes one is easier and sometimes the other: it depends on the numbers.

Method 1: using equivalent fractions

$$\frac{35}{50} = \frac{70}{100} = 70\%$$
(× 2 top and bottom)

Look at the bottom line, 50. You need to multiply it by something to make it 100. In this case you multiply by 2. Then multiply the top by the same number.

Method 2: multiplying by 100%

$$\frac{35}{50} \times 100\% = 70\%$$

One way to work this out is
$$\frac{35}{\cancel{50}_1} \times \cancel{100}^2 \% = 70\%$$

Use both methods to change each of these fractions into a percentage.

a) $\frac{9}{20}$ b) $\frac{3}{8}$

Which is the easier method in each case?

To convert from a percentage to a fraction, just remember that a percentage is a fraction with 100 on the bottom line.

Example

Convert 48% to a fraction, giving the answer in its simplest form.

$$48\% = \frac{48}{100}$$
$$= \frac{12}{25}$$

Divide both top and bottom by 4 to get the simplest form

4: Fractions and percentages

Exercise

1 For each of these, write down a percentage and a decimal equal to the fraction.

a) $\frac{8}{25}$ b) $\frac{11}{20}$ c) $\frac{1}{8}$ d) $\frac{27}{40}$ e) $\frac{179}{250}$ f) $\frac{7}{15}$

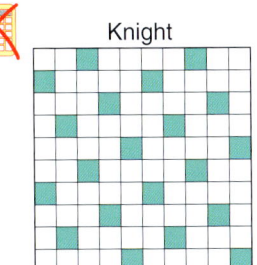

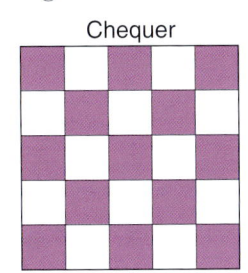

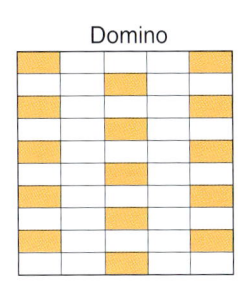

2 Look at these three patio designs.

Knight   Chequer   Domino

For each design, write down

a) the fraction of the patio that is white

b) the percentage of the patio that is white.

3 Sam's survey of 80 people found that 38 read Daily News, 25 read News Today and the rest read neither paper. Nobody reads both.

What percentage of people read

a) Daily News?   b) News Today?   c) neither paper?

4 This bar chart shows the number of male and female employees at each of a firm's two sites.

a) How many employees does the firm have altogether?

b) What percentage of the employees are male?

c) What percentage work at Parkway?

Collect five newspaper articles containing expressions like '1 in 3' or '20%'.

a) Work out each fraction as a percentage.

b) Work out each percentage as a fraction in its simplest form.

c) Explain whether you would use '44%' or '11 in 25' in a heading.

4: Fractions and percentages

Using percentages

One morning everyone in a company is given this letter.

All work stops while people work out how much they will be getting.

Asok, Helen and Fred are all paid £12 000 a year. They do the calculation differently but all get the same right answer.

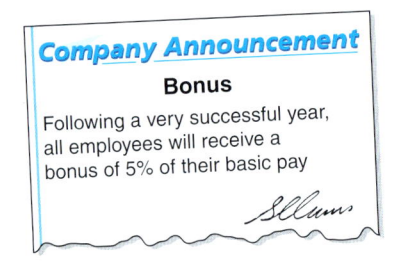

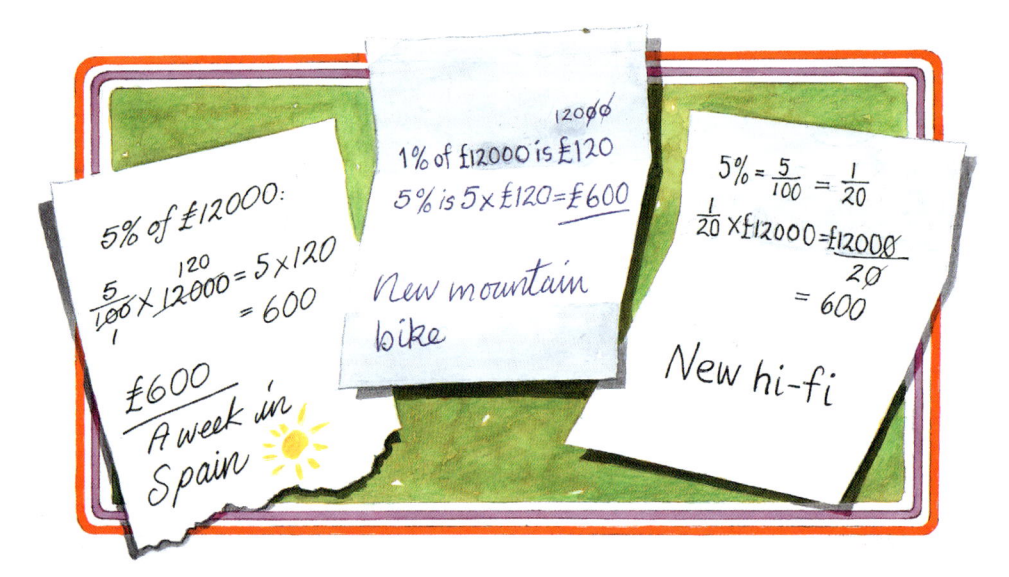

Company Announcement
Bonus
Following a very successful year, all employees will receive a bonus of 5% of their basic pay

Asok:
5% of £12000:
$\frac{5}{100} \times 12000 = \frac{5 \times 120}{1} = 600$
£600
A week in Spain

Helen:
1% of £12000 is £120
5% is 5 × £120 = £600
New mountain bike

Fred:
$5\% = \frac{5}{100} = \frac{1}{20}$
$\frac{1}{20} \times £12000 = \frac{£12000}{20} = 600$
New hi-fi

? *Look carefully at each one and make sure you can follow the working.*
Work out   a) 1% of 800   b) 25% of 40   c) 50% of 32.
Which method is easiest in each case?

Percentage change

Helen buys her new mountain bike and the next month she cycles 600 miles on it. The previous month she did only 400 miles on her old one.

This is an increase of 200 miles.

You can write it as a percentage:

$$\text{Percentage increase} = \frac{\text{increase}}{\text{original}} \times 100$$

In this case, percentage increase = $\frac{200}{400} \times 100$
= 50

Notice that a percentage increase (or decrease) is always based on the original value, not the new.

4: Fractions and percentages

1 Work out

a) 80% of 300 b) 60% of 250 c) 50% of 631

d) 15% of 580 e) $12\frac{1}{2}$% of 200 f) 3.6% of 775.

2 Emily sells her house for £75 000. The estate agent charges her 2% of the selling price.

a) How much does she pay the agent?

b) Emily later finds out that another agent is charging only 1.75%.

How much would she have saved?

3 Neil earns £8000 a year. He is given a pay rise of 3%.

a) How much does he earn after the rise?

b) The following year Neil gets a 4% pay rise.

How much does he earn after this rise?

4 This year 650 full-time and 5000 part-time students enrolled at Hatchester College. A 6% rise in full-time enrolments and a 8% rise in part-time enrolments is planned for next year. How many students in total does the college expect next year?

5 A holiday costs £400. This price is increased by 10%, then the price is reduced by 10% for last-minute bookings.

How much does a last-minute booking cost?

6 Peter currently earns £16 000 a year. He is interested in these jobs.

What percentage increase in salary would Peter get as

a) a programmer?   b) a manager?

PROGRAMMER Salary: £18 000

Manager Salary £17 200

7 Jo compares her current sales figures with those of last year.

Month	May	June	July	August
Last year	£4000	£5000	£4800	£5600
This year	£4600	£4800	£5040	£5796

a) For each month, calculate the percentage increase or decrease in sales compared with last year.

b) Calculate the percentage increase in sales during the whole four-month period.

A car costs £10 000. Each year its value decreases by 20% of its value at the start of that year.

How much is it worth
a) after one year? b) after 3 years?
c) When is it worth £$(0.8)^5 \times 10000$?

Draw a graph showing the value of the car (y axis) against its age in years (x axis). On the same graph, draw a line showing the value of a car which also costs £10 000 but which loses a fixed value of £1600 every year.

41

4: Fractions and percentages

Making comparisons

Ranjit does a survey of people's opinions of their local bus and train services. Here are his results.

Service	Number Satisfied	Number questioned
Bus	79	111
Train	37	59

? Which service is satisfying more of its customers?

Ranjit works out what proportion of the customers he asked are satisfied with each service.

$$\text{Bus: } \frac{79}{111} \qquad \text{Train: } \frac{37}{59}$$

These figures are still not easy to compare, so he writes them as percentages.

$$\text{Bus: } \frac{79}{111} = 0.711\ldots = 71\% \text{ (to nearest 1\%)}$$

$$\text{Train: } \frac{37}{59} = 0.627\ldots = 63\% \text{ (to nearest 1\%)}$$

The bus service satisfies more of its customers.

Sometimes the proportions you need to compare might be very close together. For example, which is larger, $\frac{95}{212}$ or $\frac{47}{105}$?

$$\frac{95}{212} = 0.4481132\ldots \qquad \frac{47}{105} = 0.447619\ldots$$

*The fractions give the **exact** value*

You can see from the third decimal place that $\frac{95}{212}$ is slightly larger.

Writing these numbers as whole number percentages would not show the difference – they both round to 45%.

? The only way to write $\frac{95}{212}$ or $\frac{47}{105}$ exactly is to use fractions. Why is this?

? You can write some fractions, like $\frac{1}{2}$ and $\frac{3}{10}$ exactly as decimals or percentages. What makes these special? Do the investigation on page 43 to learn more about recurring decimals.

4: Fractions and percentages

1 Write down a decimal equal to each of these fractions.

a) $\frac{4}{5}$ b) $\frac{11}{25}$ c) $\frac{3}{8}$ d) $\frac{2}{3}$ e) $\frac{1}{6}$ f) $\frac{7}{12}$

2 Arrange these numbers in order of size, starting with the smallest.

0.83 $\frac{17}{20}$ $\frac{21}{25}$ 0.09 $\frac{5}{6}$

3 Ella does a survey to find out people's opinions on two brands of personal stereo. These are her results.

Brand	Number questioned	Number reporting faults
A	129	13
B	186	21

Which brand is the more reliable?

4 Harriet manages her company's training centres. She draws up this table to compare the success of each centre.

Centre	No. of passes	No. of recruits
Northhill	35	44
Heartland	39	71
Southdown	41	52

a) Work out the percentage of recruits at each centre who passed.
b) How would you interpret these results?

5 This chart shows a manufacturer's daily output of bicycle frames from 3 production lines.

Line	Total output	Rejects
A	800	36
B	840	41
C	625	29

a) Work out the percentage of rejects for each production line.
b) Which production line is the most efficient?
c) Suggest a possible reason why the output of machine C was lower than A or B.

You know that $\frac{1}{3} = 0.33\ldots$ (going on for ever).

You say this as '0.3 recurring' and write it as $0.\dot{3}$.

Some fractions give a pattern of recurring digits.

For example, $\frac{2}{11} = 0.1818\ldots$, written $0.\dot{1}\dot{8}$, a pattern of length 2.

Investigate the patterns for $\frac{1}{7}, \frac{2}{7}, \frac{3}{7}, \frac{4}{7}, \frac{5}{7}$ and $\frac{6}{7}$.

What is the fraction with the longest pattern you can find?

4: Fractions and percentages

Finishing off

Now that you have finished this chapter you should be able to

- ★ find a fraction of a quantity
- ★ recognise simple fraction, decimal and percentage equivalents
- ★ do percentage calculations
- ★ change fractions to decimals and percentages
- ★ find proportions
- ★ make comparisons involving fractions, decimals and percentages.

Use the questions in the next exercise to check that you understand everything.

Mixed exercise

1 Work out

a) $\dfrac{7}{16} + \dfrac{3}{4}$ b) $3\dfrac{1}{4} - 1\dfrac{7}{8}$ c) $2\dfrac{3}{8} + 4\dfrac{11}{16}$ d) $1\dfrac{5}{8} + 1\dfrac{1}{2} - \dfrac{1}{4}$

2 What fraction is midway between $\dfrac{1}{2}$ and $\dfrac{3}{4}$?

3 This bar chart shows how the sales of a large company are spread across 3 countries.

a) Write down the sales (in £m) made in each country.

b) Write down the fraction of the total sales that occur in each country. Write each fraction in its simplest form.

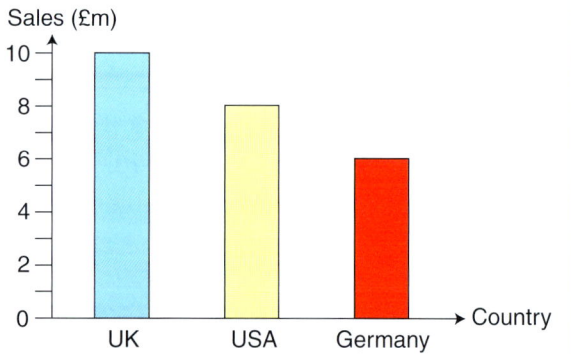

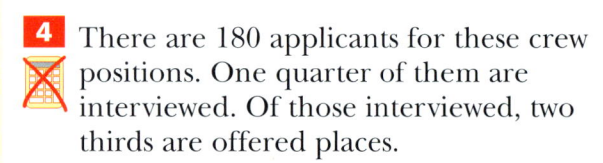

4 There are 180 applicants for these crew positions. One quarter of them are interviewed. Of those interviewed, two thirds are offered places.

a) How many of the applicants are offered places?

b) Write, in its simplest form, the fraction of the applicants who are offered places.

5 Write down the reciprocal of 6.

4: Fractions and percentages

Mixed exercise

6 Place these numbers in order of size, starting with the smallest.

$\frac{27}{8}$ $\sqrt{11}$ $\frac{10}{3}$ 3.3   3.04

7 Work out

a) 24% of 550   b) 63% of 429   c) 52.3% of 397

8 Express each fraction as a percentage.

a) $\frac{14}{25}$ b) $\frac{7}{8}$

c) $\frac{5}{6}$ d) $\frac{371}{500}$

9 The population of an island is 100 000.

It is predicted that in each year its population will rise by 5% of the figure at the start of the year.

a) Work out the predicted population after 1 year.

b) Work out the predicted population after 2 years.

10 Nita and Mark run a business. They have drawn up this table of their profits last year.

Quarter	1	2	3	4
Profit (£ thousands)	32	44	50	34

a) What was the total profit for last year?

b) What percentage of the total profit was made in the first quarter?

c) What percentage of the total profit was made in the second half of the year?

d) The expected profit for this year is £200 000. Write down the expected increase in profit as a sum of money and as a percentage of last year's profit.

11 This table shows the numbers of votes cast in 4 parish council elections.

Parish	Votes cast	No. of voters
Waterbeach	105	267
Oakington	154	309
Witchford	187	393
Northwood	137	291

Which parish had

a) the highest percentage turnout?

b) the lowest percentage turnout?

c) a turnout of approximately 2 voters in 5?

12 Ethel leaves £4800 to be divided between her grandchildren.

Ben gets a quarter of this. Jo gets 40% of the remainder.

Guy and Tanya share the rest equally.

a) How much does each grandchild get?

b) What percentage of Ethel's money does Jo get?

A snail is making a journey of 2 km. It has travelled 40 cm. Write this as a) a fraction b) a percentage of the whole journey.

What percentage of the whole journey is 1 mm?

45

Five

Area and volume

> **Before you start this chapter you should know how to work out**
>
> ★ the area of a rectangle
> ★ the area of a triangle from base and height measurements
> ★ the area of a shape made up of rectangles and triangles
> ★ the volume of a cuboid from the lengths of its sides.

Parallelograms and trapezia

Area of a parallelogram

A **parallelogram** is a shape with two pairs of parallel sides.

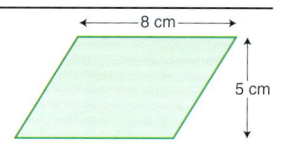

The rule for working out the area of a parallelogram is

Area of a parallelogram = base × vertical height

For this parallelogram,
  Area (in cm²) = 8 × 5 = 40

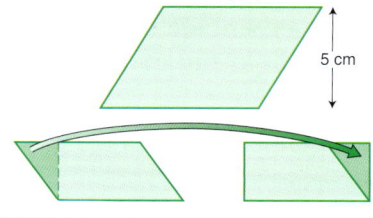

Explain why the rule works. This diagram may help.

Area of a trapezium

A **trapezium** is a shape with one pair of parallel sides.

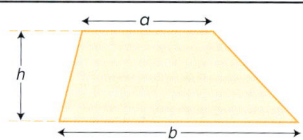

The rule for working out the area of a trapezium is

Area of a trapezium = $\frac{1}{2}(a + b)h$

For this trapezium, 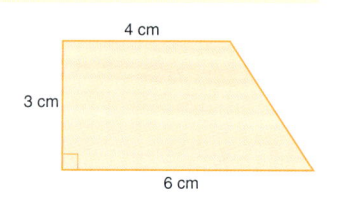
Area (in cm²) = $\frac{1}{2} \times (4 + 6) \times 3$
  = $\frac{1}{2} \times 10 \times 3$
  = $\frac{1}{2} \times 30$
  = 15

Explain why the rule works. This diagram may help. It shows two trapezia fitted together to make a parallelogram.

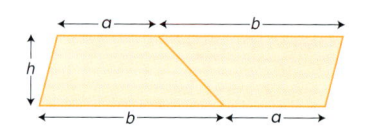

5: Area and volume

1 Find the areas of these shapes. Find the perimeters of a), b) and c).

a) 4 cm × 6 cm rectangle

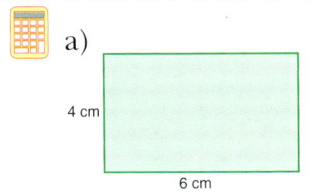

b) L-shape: 4 cm, 5 cm, 10 cm, 3 cm

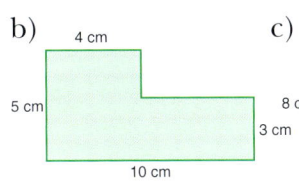

c) C-shape: 7 cm, 3 cm, 5 cm, 3 cm, 5 cm, 8 cm

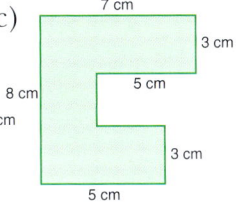

d) Triangle: base 8 cm, height 5 cm

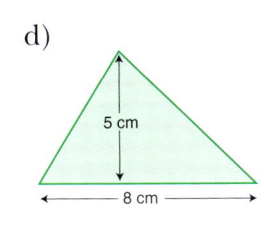

e) Kite/quadrilateral: 6 cm, 15 cm, 9 cm

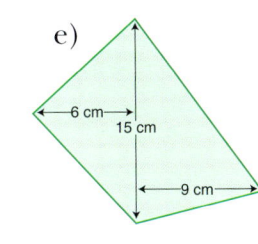

f) Pentagon-type: 4 cm, 7 cm, 12 cm, 10 cm

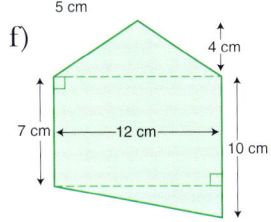

2 Find the areas of the shapes below.

a) Parallelogram: base 7 cm, height 6 cm

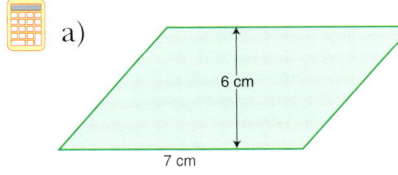

b) Trapezium: 9 cm, 14 cm, height 8 cm

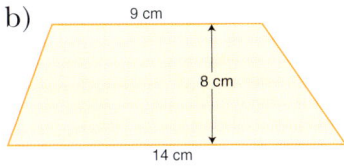

c) 7 cm, 11 cm, 7 cm

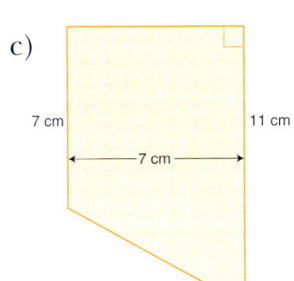

d) Parallelogram: 5 cm, 12 cm

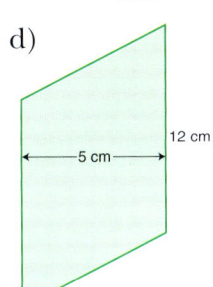

3 You are making the roof for a dolls' house out of two trapezia and two triangles, as shown.

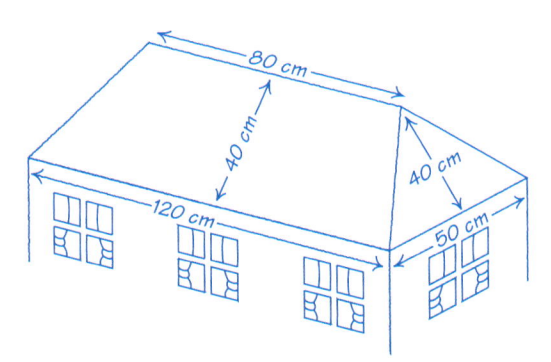

Dimensions: 80 cm, 40 cm, 120 cm, 40 cm, 50 cm

Work out the area of wood you need to make the roof.

a) Find the area of this rectangle (i) in cm^2 (ii) in m^2.

b) Use your answers to show that $1 m^2 = 100^2 cm^2$.

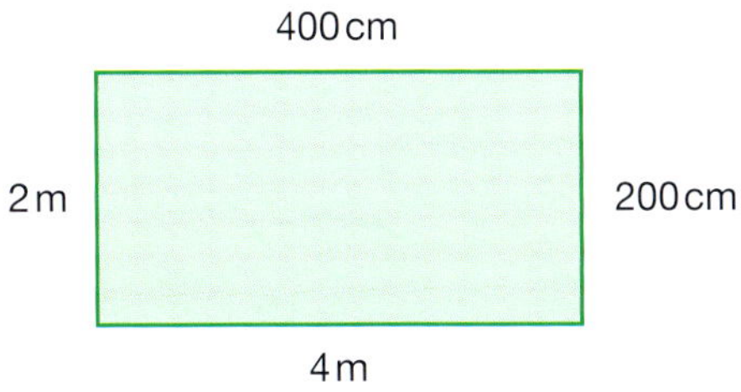

Rectangle: 400 cm / 4 m by 200 cm / 2 m

5: Area and volume

Circumference and area of a circle

The **circumference** of a circle is the distance round the edge. Look at page 3 for more information about circles.

You can find the circumference, C, and area, A, of a circle using these formulae

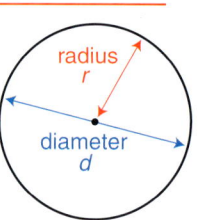

You may have a key for π on your calculator. If not, use the value 3.14 for π

$$C = \pi d$$
$$\text{or } C = 2\pi r$$
$$A = \pi r^2$$

Square the radius first, then multiply by π

Example

Find the circumference and area of a circle with diameter 6 cm.

Solution

Circumference $= \pi d$
$\qquad\qquad\quad = \pi \times 6$
$\qquad\qquad\quad = 18.85$

Area $= \pi r^2$
$\qquad = \pi \times 3^2$
$\qquad = 28.27$

The radius r is half the diameter

The circumference is 18.85 cm, and the area is 28.27 cm².

Example

Find the diameter of a circle with circumference 15 cm.

Solution

Circumference $= \pi d$
$\qquad\qquad 15 = \pi \times d$
$\qquad\qquad d = 15 \div \pi = 4.77$

To work backwards you need to do the opposite of multiplying by π. You have to divide by π

The diameter is 4.77 cm.

? Which is the exact answer, $15 \div \pi$ or 4.77?

Example

Find the radius of a circle with area 60 cm².

Solution

Area $= \pi r^2$
$\quad 60 = \pi \times r^2$
$\quad\; r^2 = 60 \div \pi = 19.1$
$\quad\;\; r = \sqrt{19.1} = 4.37$

When you work backwards you must do everything in reverse order. Divide by π first

Finding the square root is the reverse of squaring

The radius is 4.37 cm.

5: Area and volume

In this exercise first write down the exact answers, using π. Then use your calculator to work them out.

1 Find the circumference of each of these circles.

a) Diameter = 5 cm   b) Diameter = 8 cm   c) Radius = 3 cm

2 Find the area of each of these circles.

a) Radius = 2 cm   b) Radius = 7 cm   c) Diameter = 11 cm

3 Find the diameter of the circles with each of these circumferences.

a) 25 cm   b) 12 cm   c) 34 cm

4 Find the radius of the circles with each of these areas.

a) 30 cm^2 b) 56 cm^2 c) 112 cm^2

5 A label is to be wrapped round a tin which has radius 4 cm.

Find the length of the label.

6 The distance round the edge of a circular flower bed is 24 metres.

a) Find the radius of the flower bed.

b) Find the area of the flower bed.

7 Find the area of each of the following shapes. The dotted lines are there to help you.

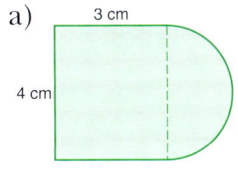

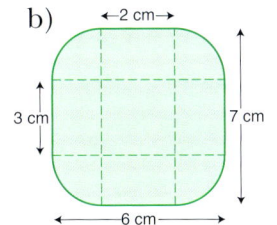

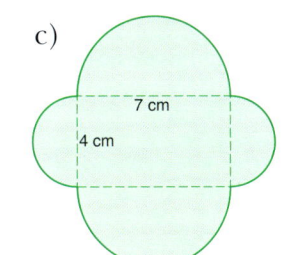

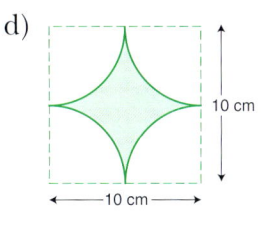

a) 3 cm, 4 cm

b) 2 cm, 3 cm, 7 cm, 6 cm

c) 7 cm, 4 cm

d) 10 cm, 10 cm

8 Find the area of

a) the quarter-circle OAB

b) the triangle OAB

c) the shaded segment.

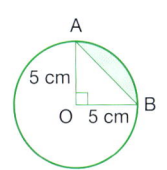

5 cm, 5 cm

Find a suitable cylinder-shaped object such as a can.
Measure its diameter, and its circumference by putting a piece of string around it. Use your results to estimate the value of π. Do you get a more accurate answer if you wind the string around several times?

The number π occurs naturally.
If you write it as a decimal it goes on for ever, and no fraction is exactly π.

Here are some approximate values for π. Which one is closest?

a) 3   b) $\frac{22}{7}$ c) $\sqrt{10}$ d) 3.14

5: Area and volume

Volume of a prism

When slices are cut (as shown) from this wedge of cheese, each piece is the same size and shape. The wedge of cheese is an example of a **prism**. A prism is a solid which has the same cross section all the way along its length.

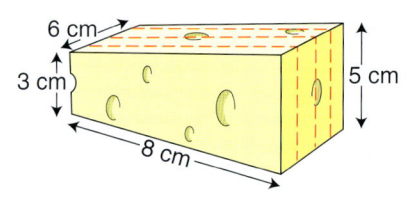

? All prisms have at least one plane of symmetry. Where is it?

? Think of 3 other familar objects whose shapes are prisms.

To find the volume of a prism, the first step is to find the area of the cross section. In the case of the wedge of cheese, this is a trapezium.

Area of trapezium (in cm^2) = $\frac{1}{2} \times (3 + 5) \times 8$
= 32

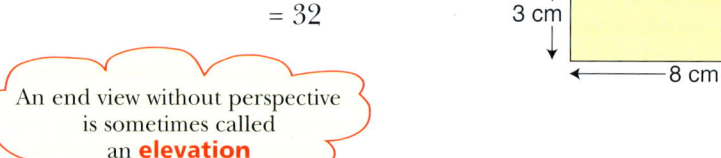

*An end view without perspective is sometimes called an **elevation***

Now multiply the area of cross section by the length.

The volume of the wedge of cheese = 32 cm^2 × 6 cm = 192 cm^3.

The volume of any prism can be found in the same way, by multiplying the area of the cross section by the length of the prism.

Volume of a prism = area of cross section × length

This Swiss roll is a prism with circular cross section. This shape is usually called a **cylinder**.

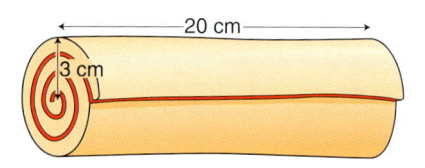

Using the above rule,

Area of cross section (in cm^2) of Swiss roll = π × 3^2 = 9π

Volume = 9π cm^2 × 20 cm = 565 cm^3 (to the nearest cm^3).

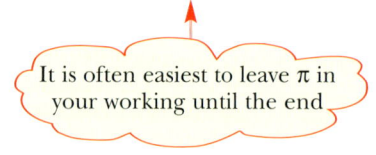

It is often easiest to leave π in your working until the end

? The formula for the volume of a cylinder is $V = \pi r^2 h$. What do V, r and h stand for?

5: Area and volume

1 A child's toy consists of five plastic prisms (shown below) with different cross sections, and a box with matching holes through which to post them.

Each prism is 5 cm long. Find the volume of each prism.

a) 3 cm × 3 cm

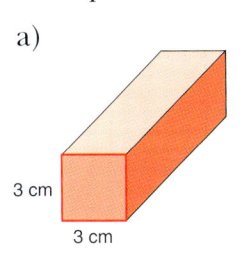

b) All sides of the cross are 2 cm

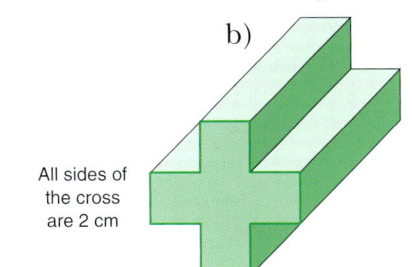

c) [cylinder]

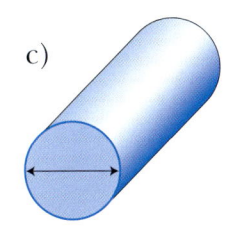

d) 3 cm, 4 cm, 5 cm (right triangle)

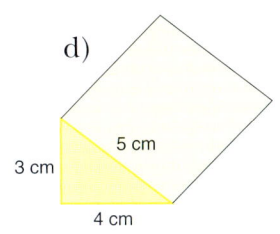

e) 3 cm (semicircle)

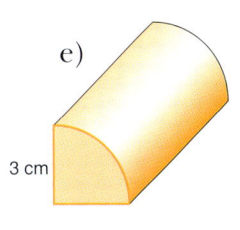

2 The cross section of a swimming pool is shown in this diagram. The pool is 10 m wide, and it is filled to the brim.

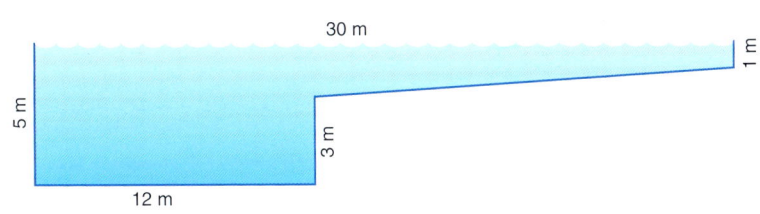

30 m, 1 m, 5 m, 3 m, 12 m

Find the volume of water in the swimming pool.

3 Metal discs for pet collars have a diameter of 4 cm. Each disc is 0.1 cm thick. How many discs can be made from 1000 cm³ of metal?

a) Find the volume of this cuboid (i) in cm³ (ii) in m³.

b) Use your answers to show that $1\,m^3 = 100^3\,cm^3$.

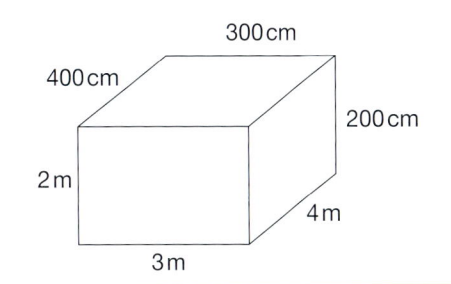

300 cm, 400 cm, 200 cm, 2 m, 3 m, 4 m

Measure the diameter and thickness of a 1p coin and a 2p coin and work out the volume of each coin. Does a 2p coin contain twice as much metal as a 1p coin?

Compare the amounts of metals in other coins, and make a table of your results.

5: Area and volume

Surface area of a prism

Fran brings home some wooden offcuts from her technology class and covers them with plastic. They will be a toy for her little sister.

One piece is a 2 cm cube.

Fran draws the **net** of this cube. The area of this net is the **surface area** of the cube. Fran needs 6 squares of side 2 cm, i.e. $6 \times 2 \times 2 = 24 \, cm^2$ of plastic.

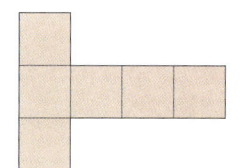

Another piece is L-shaped.

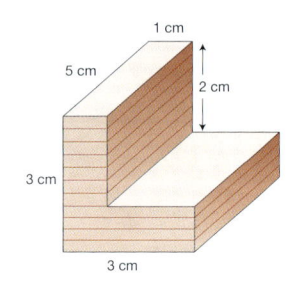

? What part of the L-shaped piece does this rectangle cover?

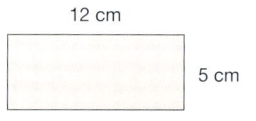

What is the complete surface area of the L-shaped piece?

Another piece is this prism. The cross section is a right-angled triangle.

? What is the area of the triangle?

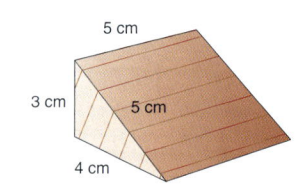

What is the perimeter of the triangle?

Fran writes this.

Ends 2 × 6 = 12
Sides 12 × 5 = 60
Total   72

? Explain Fran's work.

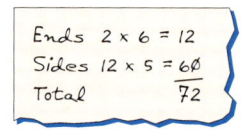

Another piece is a cylinder.

The surface area of a cylinder is
$A = 2\pi r^2 + 2\pi rh$

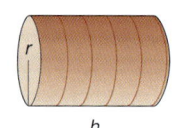

? Explain this formula.

The general formula for the surface area of a prism is

Surface area = 2 × area of cross section + perimeter of cross section × length

5: Area and volume

1 Find the surface area of each of these bricks. Each brick is 8 cm long.

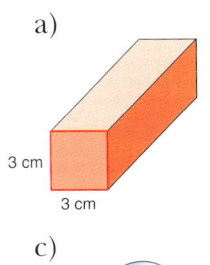

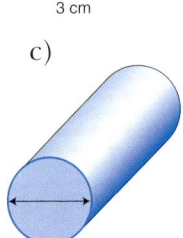

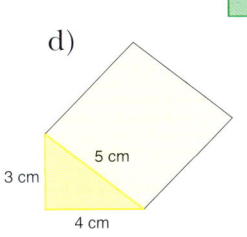

a) 3 cm, 3 cm

b) All sides of the cross are 2 cm

c) (cylinder)

d) 3 cm, 5 cm, 4 cm

e) 3 cm

2 The cross section of a swimming pool is shown in this diagram.

The pool is 10 m wide. It needs re-tiling. Find the area to be tiled.

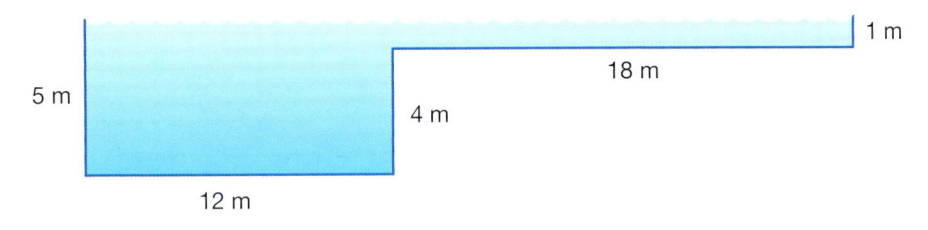

5 m, 4 m, 1 m, 18 m, 12 m

(Hint: remember that there are no tiles on the top surface.)

3 The diagram shows the cross section of a tunnel. The top is a semi-circle.

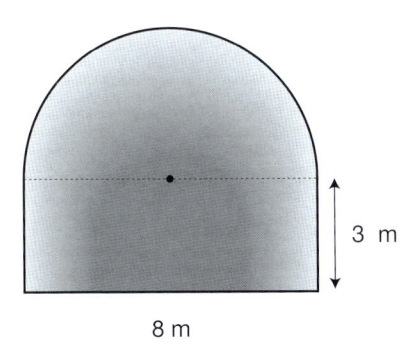

3 m, 8 m

The tunnel is 50 m long. Find the surface area of the sides and roof of the tunnel.

Here is the cross section of a shape 5 cm long.

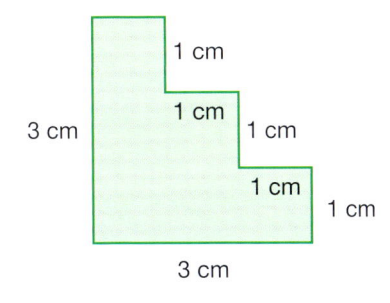

3 cm, 1 cm, 1 cm, 1 cm, 1 cm, 1 cm, 3 cm

Draw this shape on isometric paper.

Calculate its surface area and draw its net.

5: Area and volume

More about prisms

Claire designs food packaging. She is working on a new design for a packet of ground coffee.

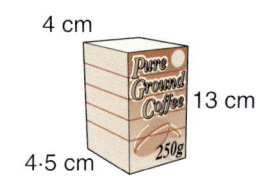

The existing packet is a cuboid as shown. The volume of this packet is

13 cm × 4.5 cm × 4 cm = 234 cm^3

The new design must have the same volume so that it holds the same amount of coffee.

Claire's first idea is a triangular prism like this:

Claire needs to work out how long the packet must be to have a volume of 234 cm^3.

Here is Claire's calculation.

First she works out the area of the cross section

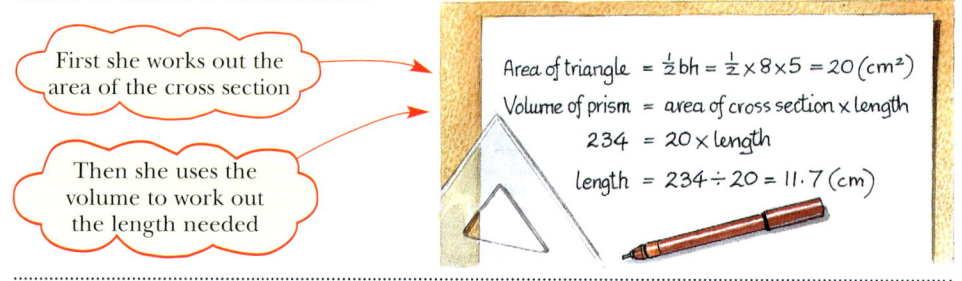

Then she uses the volume to work out the length needed

Area of triangle = $\frac{1}{2}bh = \frac{1}{2} \times 8 \times 5 = 20$ (cm^2)

Volume of prism = area of cross section × length

234 = 20 × length

length = 234 ÷ 20 = 11.7 (cm)

The packet needs to be 11.7 cm long.

Claire's next idea is a packet in the shape of a cylinder. She wants a cylinder 12 cm high. She needs to work out the radius of the cylinder.

Here is Claire's calculation.

First she uses the volume to work out the area of the cross section

Then she works out the radius of the circle

Volume of prism = area of cross section × height

234 = area of cross section × 12

Area of cross section = 234 ÷ 12 = 19.5 (cm^2)

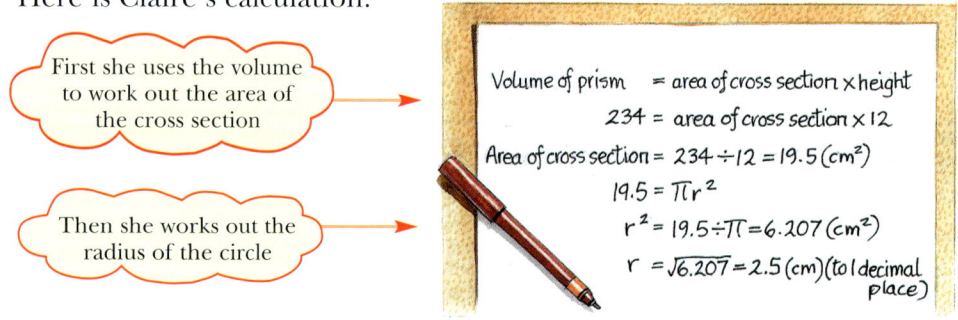

19.5 = πr^2

r^2 = 19.5 ÷ π = 6.207 (cm^2)

r = $\sqrt{6.207}$ = 2.5 (cm) (to 1 decimal place)

The radius of the cylinder should be 2.5 cm.

What are the advantages and disadvantages of Claire's designs?

5: Area and volume

1 Find the lengths marked with letters on these prisms, whose volumes are marked.

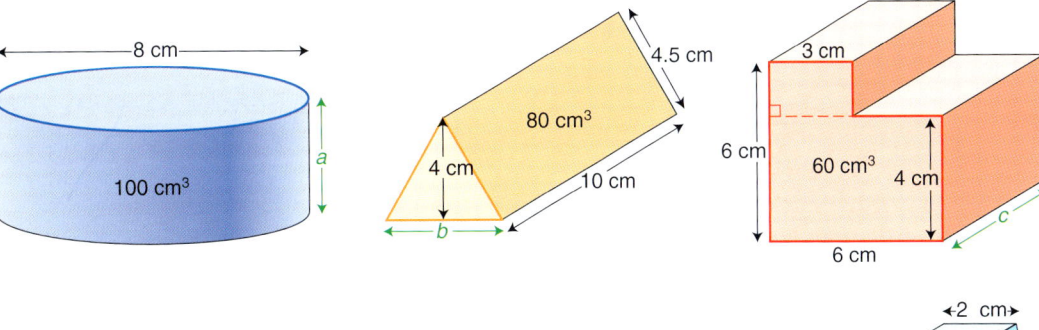

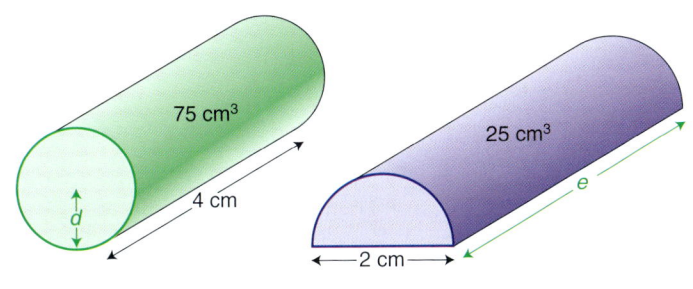

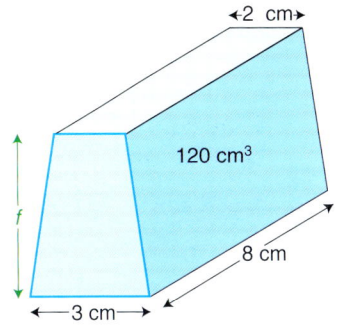

[Cylinder: diameter 8 cm, height *a*, volume 100 cm³]

[Triangular prism: triangle height 4 cm, base *b*, slant 4.5 cm, length 10 cm, volume 80 cm³]

[Step-shaped prism: 3 cm, 6 cm, 6 cm, 4 cm, length *c*, volume 60 cm³]

[Cylinder: radius *d*, length 4 cm, volume 75 cm³]

[Half-cylinder: diameter 2 cm, length *e*, volume 25 cm³]

[Trapezoidal prism: top 2 cm, bottom 3 cm, height *f*, length 8 cm, volume 120 cm³]

2 Find the surface area of the first five prisms above.

3 A drinks company is trying out different sizes for cans of fizzy drink.

a) One size is to hold 250 ml. The radius of the base of the can is 3 cm. Find the height of the can.

b) Another size is to hold 400 ml. The height of the can is 12 cm. Find the radius of the base.

4 A skip has a capacity of 6 m³. It is in the shape of a prism with the cross section shown. How wide is the skip?

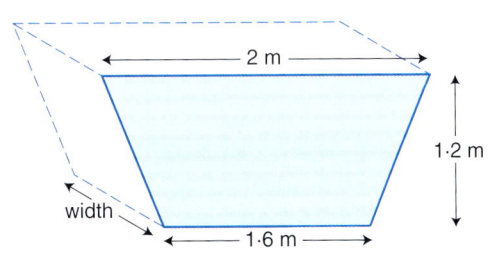

[Trapezoidal cross section: top 2 m, bottom 1·6 m, height 1·2 m]

Find 4 containers, each in the shape of a prism.

For each one, draw a diagram and mark on its dimensions. Work out the capacity of each container.

5: Area and volume

Using dimensions

You have now met quite a lot of different formulae for lengths, areas and volumes. There are many more! Some of them are quite similar and it can be difficult to remember which is which. It is possible to tell just by looking at a formula whether it is

$2\pi r$ $\pi r^2 h$
lbh $2l+2h$
$\frac{1}{2}(a+b)h$ πr^2

for a length, an area or a volume. The way you can tell is by deciding how many **dimensions** there are in it. This means the number of length measurements that have been multiplied together. Numbers, like 2 or π or $\frac{1}{2}$, don't count, because they aren't measurements of length.

- Lengths have one dimension
- Areas have two dimensions
- Volumes have three dimensions.

Example 1

lbh

l, b and h are all measurements of length. The three measurements are multiplied together

The formula has **three** dimensions. So this formula is a **volume**.

? *What shape has the formula Volume = lbh?*

Example 2

π is not a measurement so it is not a dimension

πr^2

r^2 counts as two dimensions as it means $r \times r$

The formula has two dimensions. So this formula is an area.

? *What shape has the formula Area = πr^2?*

Example 3

In some formulae there are two or more parts added together. Each part must have the same number of dimensions. The number of dimensions of the whole formula is the same as the number of dimensions of each part.

$2l$ has one dimension

$2l+2h$

$2h$ has one dimension

The formula has one dimension. So this formula is a length.

? *What shape has the formula Perimeter = $2l + 2h$?*

5: Area and volume

1 Each of these formulae is either a length (perimeter or circumference), area or volume of one of the shapes shown below. For each formula:

a) say how many dimensions it has and what they are
 (Example: $3a^2b$ has three dimensions, a, a and b)

b) say whether the formula is a length, area or volume

c) say which of the shapes A – F it belongs to.

(i) x^2y (ii) $2(x+y)$ (iii) πy^2 (iv) xy

(v) $\frac{1}{2}y(x+y)$ (vi) $\pi x^2 y$ (vii) $\frac{1}{2}xy$ (viii) $2\pi y$

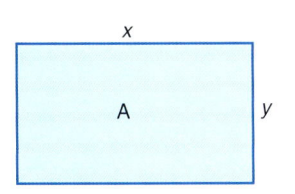

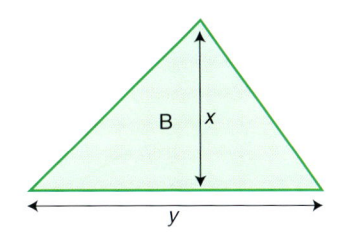

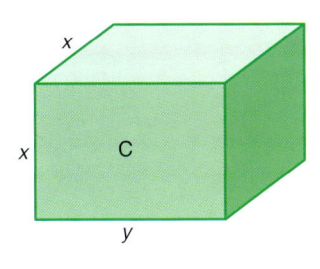

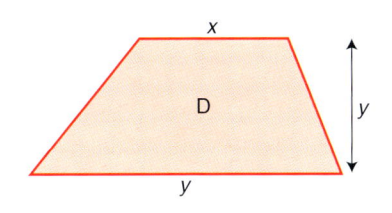

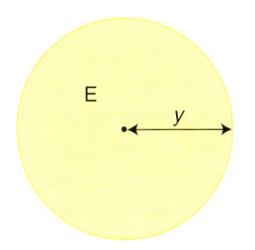

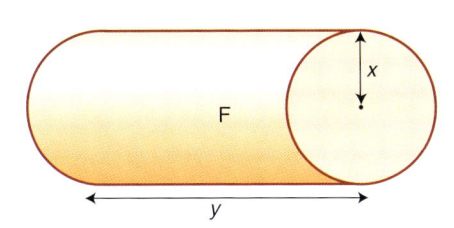

2 Some of the formulae below are real formulae for length, area or volume. Some could not be real formulae because they have the wrong number of dimensions. For each formula, write down whether it is a length, an area, a volume or not a real formula. r, h, a and b are all measurements of length. Any other letters used are not.

a) $\frac{1}{3}\pi r^2 h$ b) $2(ah+bh+ab)$ c) $4a+2b$ d) $2\pi rh + 2\pi r^2$

e) $\pi r^2 a^2$ f) $\frac{1}{2}ab\sin C$ g) $2r^2 + a^2 b$ h) $\frac{4}{3}\pi r^3$

5: Area and volume

Finishing off

Now that you have finished this chapter you should be able to

★ find the circumference of a circle

★ find the area of a circle

★ find the area of a parallelogram

★ find the area of a trapezium

★ find the volume and surface area of a prism

★ find other dimensions of prisms from the volume

★ distinguish between formulae for perimeter, area and volume by considering dimensions.

Use the questions in the next exercise to check that you understand everything.

Mixed exercise

1 Find the area of each of these shapes.

a) [trapezium with parallel sides, 4 cm and 6 cm on the sides, 10 cm across]

b) [parallelogram with height 6 cm and base 9 cm]

2 Find the area and perimeter of each of these shapes.

a) [circle with diameter 12 cm]

b) [shape: rectangle with semicircular ends, 6 cm length of rectangle, 2 cm radius]

3 Find the volume and surface area of each of these prisms.

a) [step-shaped prism: 3 cm, 4 cm, 2 cm, 5 cm, 6 cm]

b) [rectangular prism: 5 cm, 8 cm, 5 cm, 10 cm, 4 cm]

c) [cylinder: radius 4 cm, length 12 cm]

58

5: Area and volume

Mixed exercise

4 A baby's bottle is approximately the shape of a cylinder. The diameter of its base is 6 cm. The bottle can be filled to a maximum depth of 9 cm.

a) What is the maximum amount of milk that the bottle can hold?

b) Baby Sam drinks 150 ml of milk at each feed (1 ml is the same as $1\,cm^3$).
To what depth should his bottle be filled?

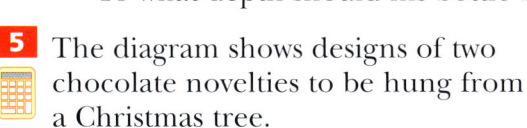

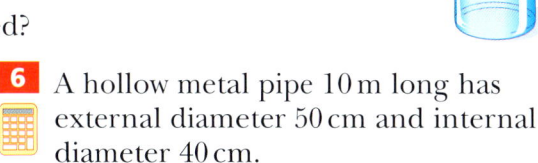

5 The diagram shows designs of two chocolate novelties to be hung from a Christmas tree.

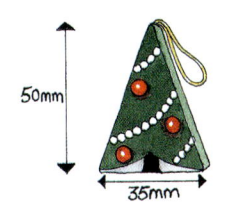

a) The 'Christmas tree' shape is 6 mm thick.

Find the volume of chocolate in the 'Christmas tree' shape.

b) The 'Santa' shape is 5 mm thick. It must have the same volume of chocolate as the 'Christmas tree'.

Find the diameter of the 'Santa' shape.

6 A hollow metal pipe 10 m long has external diameter 50 cm and internal diameter 40 cm.

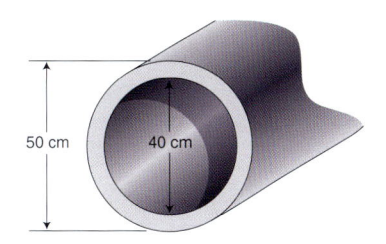

Find the volume of metal needed to make the pipe.

(Hint: find the volume of a solid cylinder and then subtract the volume of the hole.)

7 A carton of fruit juice contains one litre of juice.

Remember that 1 litre = $1000\,cm^3$

The carton is 16 cm long and 5 cm wide.

How deep is the juice in the carton?

8 Name the lines and regions marked with letters in these diagrams.

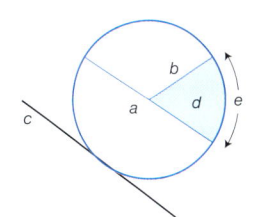

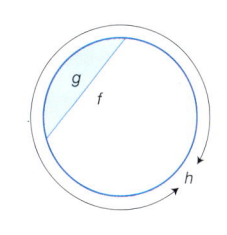

9 For each formula, say whether it is a formula for length, area, volume or none of these. The letters p, q, r and h are all measurements of length.

a) $3pq^2$ b) $\pi r h$

c) $pr^2 h$ d) $\pi(p+q)$

e) $p^2 r + qh$ f) πr^3

g) $r(p+q)$

Design several different shaped boxes to hold 6 golf balls. Compare the amounts of card needed to make each box, and the amount of space each box takes up. What other considerations would you need to take into account to choose the best design for a box?

Six

Using symbols

Before you start this chapter you should

★ be familiar with the work in Chapter 3.

Being brief

This station sign is the longest in Britain. It takes some effort to say the full name so most people call it Llanfair PG.

In maths we often need to find ways to write things briefly. For example:

$2 \times 2 \times 2 \times 2 \times 2 \times 2 \times 2$ is usually written as 2^7.

There are seven 2s here

You say this as '2 to the power 7'. The 7 is called the **power** or **index**.

Indices is the plural of index

Indices are useful in algebra too. For example:

$a \times a \times a \times a = a^4$ and $5a^4 = 5 \times a \times a \times a \times a$

Notice there is only one 5 here

How would you simplify $a^5 \times a^3$?

$a^5 \times a^3 = a \times a \times a \times a \times a \quad \times \quad a \times a \times a$
$\ 1\ \ 2\ \ 3\ \ 4\ \ 5 \qquad\qquad 1\ \ 2\ \ 3$

5 + 3 = 8. You have added the indices

$= a^8$

? Work out 2^5, 2^3 and 2^8 and so check that $2^5 \times 2^3 = 2^8$.

Here are two expressions that can be written more briefly:

$4 \times x^3 \times 3$ \qquad and \qquad $4x \times 3x^2$

$= 4 \times 3 \times x^3$ \qquad\qquad\qquad $= 4 \times 3 \times x^1 \times x^2$

$= 12x^3$ \qquad\qquad\qquad\qquad $= 12x^3$

x can be written as x^1

They turn out to be the same

Add the indices: 1 + 2 = 3

How do you simplify $a^5 \div a^3$?

5 − 3 = 2

$a^5 \div a^3 = \dfrac{a \times a \times \cancel{a} \times \cancel{a} \times \cancel{a}}{\cancel{a} \times \cancel{a} \times \cancel{a}} = a^2$

Cancelling the as leaves 1 in the denominator

? What is $a^{106} \div a^{102}$?

What rule did you use?

6: Using symbols

1 Write each of these using powers.

a) $5 \times 5 \times 5$
b) $8 \times 8 \times 8 \times 8 \times 8$
c) $10 \times 10 \times 10$
d) $7 \times 7 \times 7 \times 7 \times 7 \times 7 \times 7 \times 7 \times 7$
e) $100\,000\,000$
f) 10
g) $x \times x \times x \times x$
h) $4 \times y \times y$
i) $9 \times n \times n \times n$
j) $3 \times a \times a \times a \times a \times a$

2 Write each of these out in full.

a) 2^3
b) 6^4
c) d^3
d) n^5
e) $5c^2$
f) $7g^4$
g) $10z^1$
h) $2x \times x^5$
i) $m^2 \times 2m^3$
j) $4u^2 \times 2u$

3 Write these as briefly as possible using powers.

a) $3^2 \times 3^2$
b) $6^7 \times 6^3$
c) $4^3 \times 4^5$
d) $10^6 \times 10^{11}$
e) $a^2 \times a^3$
f) $v^8 \times v^3$
g) $s \times s^4$
h) $d \times d^7$
i) $x^2 \times x^2 \times x^3$
j) $n^3 \times n \times n^4$

4 Write these as briefly as possible using powers.

a) $2x \times 3$
b) $4x^2 \times 5$
c) $2a^3 \times 3$
d) $2 \times 3g^2$
e) $2x \times 3x$
f) $4x \times 5x$
g) $3p \times p^2$
h) $y^4 \times 7y$
i) $2n \times n \times n^2$
j) $9m \times 10m^2 \times m^3$

5 Which is biggest, 7^2, 2^7 or 2×7?

Work them all out and write them in order of increasing size.

6 Which is the biggest number? Put them in order of increasing size.

1^{10}, 2^9, 3^8, 4^7, 5^6, 6^5, 7^4, 8^3, 9^2, 10^1

When you have done this, use a calculator to check your list.

7 Work these out and write the answers using powers.

a) $3^5 \div 3^3$
b) $2^{10} \div 2^4$
c) $5^{99} \div 5^{94}$
d) $x^8 \div x^8$

8 Find the value of $x^2 + 4$ when

a) $x = 2$
b) $x = 10$
c) $x = 0$

9 Find the value of $2x^3$ when

a) $x = 2$
b) $x = 10$
c) $x = 0$

A kilometre is 1000 metres but a kilobyte is 1024 bytes.

Explain how these 2 numbers arise.

How many bytes are there in a Megabyte and a Gigabyte?

6: Using symbols

Using negative numbers

? By how many degrees did the temperature rise?

A change in temperature is given by

   new temperature – old temperature.

In this case, the change (in degrees Celsius) is $12 - (-8)$.

But what is the answer to $12 - (-8)$?

Look at the thermometer.

You can see by counting that the change from -8 °C to $+12$ °C is $+20$ °C.

So $12 - (-8) = 20 = 12 + 8$.

You can see that $-(-8) = +8$.

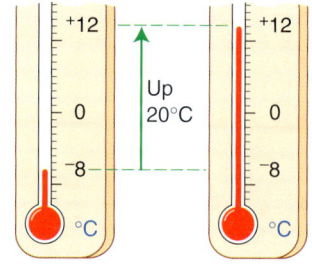

Where there are two signs before one number, the signs follow these rules.

   $+ (+8) = +8$ $+ (-8) = -8$ $- (+8) = -8$ $- (-8) = +8$

? On the Tuesday of the same week in Minneapolis, the temperature was -13 °C.

Use the rules above to work out how much the temperature rose to get to -8 °C.

Sketch a thermometer and count degrees to check your answer.

? Work these out and check your answers on a thermometer scale or a number line.

(If you get a negative answer that means the temperature has dropped.)

Old temperature = 20 °C, new temperature = 6 °C. What is the change?

Old temperature = –20 °C, new temperature = –6 °C. What is the change?

Old temperature = 6 °C, new temperature = –6 °C. What is the change?

Old temperature = 20 °C, change = –6 °C. What is the new temperature?

In algebra you often need to substitute negative numbers into expressions. The same rules apply.

Example

Work out the value of $a - b$ when $a = 25$ and $b = -8$.

Solution

$a + b = 25 - (-8)$

$\quad\quad = 25 + 8$

$\quad\quad = 33$

When a is 25 and b is –8, $a - b = 33$.

6: Using symbols

Copy each expression and work down the page when you have to do more than one calculation.

1 Simplify each of these.
   a) $-(+5)$
   b) $+(+5)$
   c) $+(-5)$
   d) $-(-5)$
   e) $-(+7)$
   f) $-(-3)$
   g) $+(-12r)$
   h) $-(+14s)$
   i) $-(-99t)$

2 Work these out.
   a) $10 - (+5)$
   b) $12 + (+5)$
   c) $-16 + (-5)$
   d) $8 - (-5)$
   e) $2 - (+7)$
   f) $5 - (-3)$
   g) $48 + (-12r)$ when $r = 2$
   h) $10 - (+14s)$ when $s = 2$
   i) $900 - (-99t)$ when $t = 1$
   j) $9 + (-3x)$ when $x = 3$

3 Write these as simply as possible
   a) $9 - (-m)$
   b) $9 - (+m)$
   c) $3 + (-a)$
   d) $3 - (-2a)$
   e) $4 + (-2) \times n$
   f) $7 - (-2) \times n$

4 Work out the value of $T - Y$ when
   a) $T = 6$ and $Y = 2$
   b) $T = -6$ and $Y = 2$
   c) $T = 6$ and $Y = -2$
   d) $T = -6$ and $Y = -2$

5 A train leaves London on Tuesday evening for Plymouth. The journey time should be 3 hours 35 minutes. In fact the train arrives 4 minutes early, at 0006 on Wednesday morning.

What time does the train leave London?

6 Find the value of $3x + 4$ when
   a) $x = 1$
   b) $x = -1$
   c) $x = 2$
   d) $x = -2$

Work out the value of
a) $-25 + 49$
b) $+25 - 49$
c) $+25 + 49$
d) $-25 - 49$

Use your results to help you write down simple rules about how to add and subtract any 2 numbers, positive or negative.

6: Using symbols

Simplifying expressions with negative numbers

What happens when you multiply a negative number by a positive one, for example $(-3) \times 2$?

You can see this on the number line. Multiplying by 2 doubles the distance from zero.

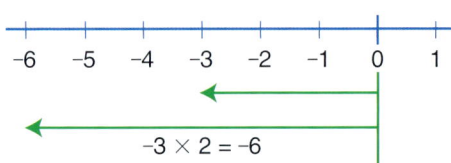

$-3 \times 2 = -6$

You can write this the other way round too, as $2 \times (-3) = (-6)$.

In this, 2 means (+2), so you can see that

$- \times + \rightarrow -$

and

$+ \times - \rightarrow -$

What happens when two negative numbers are multiplied together? What is $(-2) \times (-2)$?

This example gives you the answer. Sam goes shopping with £15. He buys two of these T shirts. Together they cost £10 so he has £5 left. You can also write it down like this.

Money left = money at start − cost of 2 T-shirts

In £: $5 = 15 - 2 \times (7 - 2)$

$5 = 15 - 2 \times 7 - 2 \times (-2)$ ← expand the bracket

$2 \times 7 = 14$

$5 = 15 - 14 - 2 \times (-2)$

$5 = 1 - 2 \times (-2)$

$15 - 14 = 1$

But $5 = 1 + 4$ so this shows you that $(-2) \times (-2) = (+4)$.

When you multiply two negative numbers together the answer is positive.

$- \times - \rightarrow +$

$+ \times + \rightarrow +$

This is the same for positive numbers.

What is the value of $(-5) \times (-5)$?

Why is the square root of 25 written as ± 5?

What is the positive square root of 25? What is the negative square root of 25?

64

6: Using symbols

Exercise

1 Use the rules on page 64 to work these out.

a) $4 \times (-2)$ b) -4×2 c) $-4 \times (-2)$ d) $+6 \times (+4)$

e) $+8 \times (-3)$ f) $-5 \times (-4)$ g) $-3 \times (-3)$ h) $(-8)^2$

i) $-30 \div 5$ j) $-30 \div (-5)$ k) $30 \div (-5)$ l) $55 \div (-5)$

m) $-2 \times (5x)$ n) $-4 \times (-2x)$ o) $-3x \div (-1)$

2 Simplify each of these. Remember to carry out the operations in the right order.

a) $2 \times (-3) + 15$ b) $5 - 2 \times (-3)$ c) $5 - 2 \times (+3)$

d) $3 + 12 \div (-4)$ e) $15 \div (-3) + 6$ f) $17 - 6 \div (-3)$

3 Expand these brackets.

a) $4(2z + 7)$ b) $4(2z - 7)$

c) $-4(2z + 7)$ d) $-4(2z - 7)$

e) $10(6 + 2a)$ f) $-10(6 + 2a)$

g) $-10(6 - 2a)$ h) $7(-e - 2)$

i) $-(4 - 8y)$ j) $-(4 + 8y)$

4 Expand the brackets in each of these, then simplify them. Check your answers.

a) $12 + 2(5 - 3)$ b) $12 - 2(5 - 3)$

c) $12 - 2(5 + 3)$ d) $14 - (7 + 4)$

e) $14 + (7 - 4)$ f) $14 - (7 - 4)$

g) $3(5 - 4) - 6$ h) $(2 + 8) + 2(1 - 3)$

5 Expand the brackets then simplify these.

a) $21 + 2(x + 3)$ b) $21 - 2(x + 3)$ c) $6n + (4n - 10)$

d) $6n - (4n - 10)$ e) $5d + 4(1 + 2d)$ f) $19a + 3(10b - 6a)$

g) $3(5 - 3t) - t$ h) $(2 + 8c) + 2(1 - c)$ i) $23 - (5 + 4p) + 4p$

6 Find the value of $4(p - 3q) - (7 - q)$ when

a) $p = 2$ and $q = 1$ b) $p = 2$ and $q = -1$ c) $p = -2$ and $q = 5$

Investigation

What happens to the sign when you keep multiplying negative numbers?

Write down the value of $(-1)^2$, $(-1)^3$, $(-1)^4$, and so on.

What is the value of $(-1)^{213}$?

Find out how to do calculations like $(-5) \times (-4)$ on your calculator.

Write some instructions to enable a friend to do them.

6: Using symbols

Finishing off

Now that you have finished this chapter you should be able to

★ multiply and divide powers of a number

★ multiply and divide positive and negative numbers

★ use a number line to add and subtract positive and negative numbers

★ expand brackets that are multiplied by a negative number.

Use the questions in the next exercise to check that you understand everything.

Mixed exercise

1 Write each of these as briefly as you can.

a) $4 \times 4 \times 4 \times 4 \times 4$
b) $y \times y \times y$
c) $5 \times x \times x$
d) $7 \times a \times a \times a$
e) $5 \times b \times b \times 3 \times b$
f) $x^2 \times x^5$
g) $3y^3 \div y^2$
h) $15p^8 \div 5p^3$
i) $4m^2 \div 4m$
j) $d^4 \div 4d^3$

2 Simplify each of these.

a) $6 \times (-3)$
b) $-4 \times (-2)$
c) $12 \div (-4)$
d) $-12 \div 4$
e) $(-12) \div (-4)$
f) $8 + 3 \times (-2)$
g) $-24 \div 2 + 6 \times 5$
h) $3 \times 4x - 2 \times 2x$
i) $6 + (-2)$
j) $-8 - (-5)$
k) $3y - (-2y)$
l) $-4m + (-2m)$

3 Simplify these.

a) $5t^2 + 4t^2$
b) $8s^2 - 3s^2$
c) $7r^2 - r^2$
d) $9q^2 - 4q^2 + 2q^2$
e) $p^2 + 2p^2$
f) $n^2 - 4n^2 + 8n^2$

4 Work out the value of

a) $2 + x$ when $x = -3$
b) $4 - m$ when $m = -6$
c) $a - (-a)$ when $a = -1$
d) $n + n - 4$ when $n = -2$.

6: Using symbols

Mixed exercise

5 In each of these, expand the brackets and simplify the answer.

a) $2(a + 3) + a$
b) $5b - 2(1 + b)$
c) $3(2c - 2) - 5c$
d) $3(d + 2) - 2(d + 3)$
e) $12e - 6(4 + 2e)$
f) $4(2 + f) - 3(1 - f)$
g) $(3g + 1) - (2g + 4)$
h) $5h - (3h + 2)$
i) $4(a + b) - 3(a - b)$
j) $8a + 4(b - 2a) - 5b$
k) $3(2x + y - 4z)$
l) $4(x + y) - 3(x + y) - (x + y)$

6 Rupal went shopping with £r, and bought 3 presents for £p each and 3 sheets of wrapping paper at £q each.

Write down an expression for the number of £ she had left.

Check that this gives the right answer in the case when $r = 20$, $p = 4$ and $q = 0.5$.

7 Amber went shopping with £a. She returned a skirt costing £b to a shop, and bought another costing £c.

Write down an expression for the number of £ Amber had left.

Check that your expression gives the right answer in the case when $a = 25$, $b = 17$ and $c = 21$.

8 Find the value of $3x^2 + 4$ when

a) $x = 1$
b) $x = -1$
c) $x = 0$
d) $x = 4$
e) $x = -4$

Investigation

Make ten cards with numbers and signs on like this:

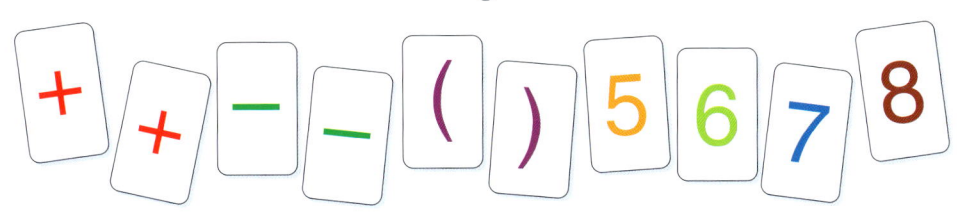

$+$ $\times$ $-$ $-$ $($ $)$ 5 6 7 8

Using the numbers as single digits (i.e. not as 56, 75 ...), combine some or all of the cards to make different expressions.

What are the largest and smallest numbers you can make?

Write down as many combinations as you can which give the answers
a) 0   b) 14.

Seven

Data handling

Before you start this chapter you should

★ understand the difference between categorical and numerical data

★ understand the difference between discrete and continuous data

★ be able to make a tally chart

★ be able to draw up a frequency table.

Pie charts

Each year, many wild animals are killed on the roads. The members of a conservation group keep a record of the bodies they see on the roads in one county.

This table shows the types of road on which these animals died.

Road type	Motorways	A roads	B roads	Other	Total
Deaths	120	1040	560	1160	2880

Mick is designing a pie chart to illustrate these data.

First he works out how many casualties will be represented by 1° on his chart.

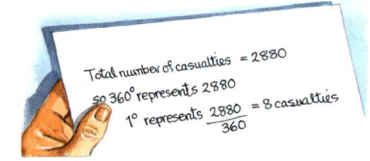

Then he works out the angle that each type of road needs on the chart. Here is his working for motorways.

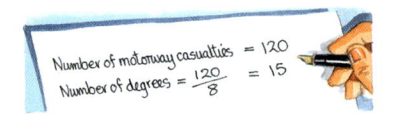

Total number of casualties = 2880
so 360° represents 2880
1° represents $\frac{2880}{360}$ = 8 casualties

Number of motorway casualties = 120
Number of degrees = $\frac{120}{8}$ = 15

When he has worked out all the angles, Mick draws his chart.

Each 'slice of pie' is called a **sector**

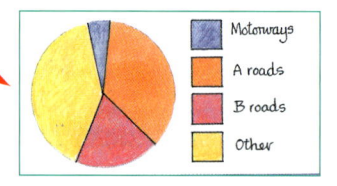

What does Mick's chart tell you?

Is it what you would have expected?

Often, rather than drawing a pie chart yourself, you need to read information from one that someone else has drawn. The chart illustrates the types of animals involved in the 120 deaths on motorways.

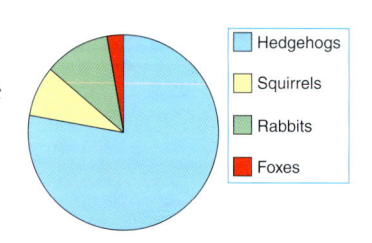

Measure the angle for hedgehogs and work out how many this represents.

7: Data handling

Exercise

1 A senior citizens' group is campaigning for better state pensions. Alice (who is 75 and lives alone) allows them to use her finances as an illustration. This list shows how she spends her £72 weekly pension.

Draw a pie chart to show this information.

Rent	£12
Gas/electricity	£9
Food	£22
Household	£12
Personal	£6
Clothing	£4
Fares	£3
Other	£4

2 Of 1440 business students, 576 started careers in Finance, 360 in Sales, 288 in Management, 108 in Administration, 72 in Market Research and the rest in Marketing.

Calculate the angle required for each sector and draw a pie chart to illustrate these data.

3 This pie chart shows the holiday destinations of a number of passengers at Heathrow Airport.

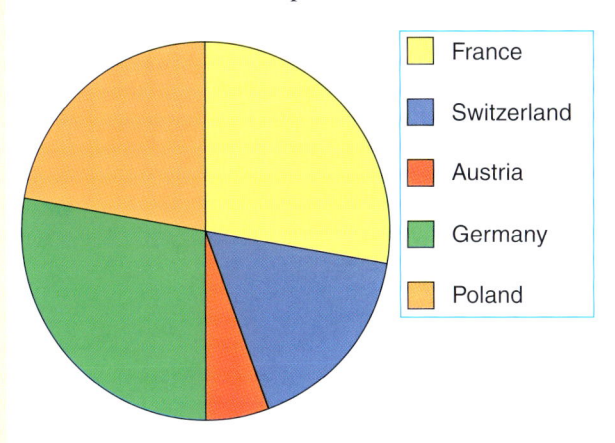

- France
- Switzerland
- Austria
- Germany
- Poland

a) The number of people going to Poland is 40. How many passengers are involved altogether?

b) How many people are travelling to each destination?

Sometimes the data you want to show in a pie chart may be given as percentages. If so, a pie chart scale (an angle measure marked in percentages) makes it very easy to draw the chart.

Find out the percentages of the different gases in air. Use a pie chart scale to draw a pie chart illustrating this.

Using road safety data for your local area, draw an eye-catching poster, including a pie chart, to inform people about when or where they are most at risk.

7: Data handling

Stem-and-leaf diagrams

Mel and Jane measure the heights of all the girls in their class to the nearest centimetre and record the results.

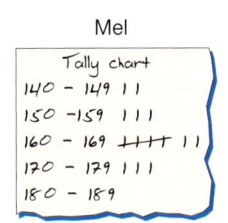

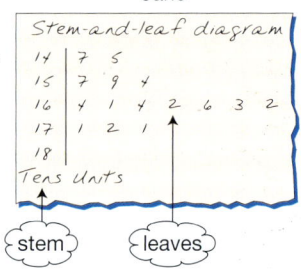

Mel — Tally chart
- 140 – 149 ||
- 150 – 159 |||
- 160 – 169 |||| ||
- 170 – 179 |||
- 180 – 189

Jane — Stem-and-leaf diagram

```
14 | 7 5
15 | 7 9 4
16 | 4 1 4 2 6 3 2
17 | 1 2 1
18 |
```
Tens Units — stem / leaves

? What advantages does Jane's stem-and-leaf diagram have over Mel's tally chart?

Are there any advantages in Mel's tally chart?

Jane orders her stem-and-leaf diagram and adds a key to explain it.

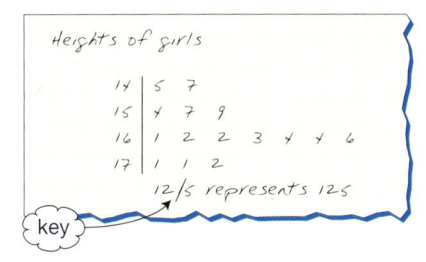

Heights of girls
```
14 | 5 7
15 | 4 7 9
16 | 1 2 2 3 4 4 6
17 | 1 1 2
```
12/5 represents 125 (key)

? How many girls are 164 cm?

Marion's height is 159 cm. Find the entry corresponding to Marion.

What is the smallest height?

How tall is the tallest girl?

Mel and Jane also measure the boys and Jane draws a back-to-back diagram to compare the two sets of data. She puts the number of leaves beside the diagram.

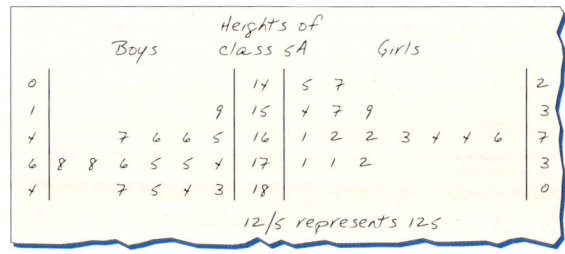

Heights of class 5A

Boys			Girls	
0		14	5 7	2
1		15	4 7 9	3
4	7 6 6 5	16	1 2 2 3 4 4 6	7
6	8 8 6 5 5 4	17	1 1 2	3
4	7 5 4 3	18		0

12/5 represents 125

? How tall is the tallest student in the class?

Allie is 166 cm. How many girls are taller than Allie? How many boys are taller than Allie?

Do you think the boys are taller than the girls?

The students are divided into five groups according to their heights.

? Which group is the largest? This is called the **modal group**.

70

7: Data handling

1 A baker keeps a record of the number of sponge cakes he sells each day.

30 35 61 73 64 62 59 33 42 55
34 36 45 42 39 51 47 38 42 30
43 45 65 53 57 42 45 34 37 65

a) Construct a sorted stem-and-leaf diagram to show the data.
b) What is the smallest number sold?
c) What is the greatest number sold?
d) On how many days are 45 sponges sold?

2 The numbers of miles travelled by students to a conference were as follows.

23 46 63 25 39 47 52 61 69 58
47 53 47 45 63 47 59 47 65 33
29 47 35 47 68 21 25 57 47 47

a) Construct a sorted stem-and-leaf diagram to show the data.
b) What was the smallest number of miles travelled?
c) What was the greatest number of miles travelled?
d) How many students travelled 47 miles?

3 The ages of skaters at a local skating rink are recorded in a survey.

7 24 27 17 9 11 13 25 22 18
37 27 14 17 23 31 16 8 19 6
23 15 14 27 34 26 17 13 14 14

a) Construct a sorted stem-and-leaf diagram to show the data.
b) How old is the youngest skater?
c) How old is the oldest skater?
d) How many 17-year-olds are present?
e) What is the most common age?

4 Jane and Robert live next door to each other. They both planted 20 hollyhocks. Jane used a fertiliser but Robert didn't. The heights, in centimetres, of their hollyhocks were as follows.

Jane: 185 201 256 248 200 254 239 234 196 223
      199 243 257 239 246 226 229 254 180 250

Robert: 208 245 150 228 230 215 217 228 215 158
        164 179 208 212 226 230 188 196 214 226

a) Construct a sorted back-to-back stem-and-leaf diagram to show the data.
b) State the heights of the tallest and shortest hollyhocks belonging to Jane.
c) State the heights of the tallest and shortest hollyhocks belonging to Robert.
d) Compare the two sets of data and comment on whether the fertiliser is effective.

Construct a stem-and-leaf diagram with the following features.
- The range from lowest to highest value is 50.
- The number of values is 15.
- The highest value is 89.
- The middle value is 53.

7: Data handling

Moving averages

Dennis has kept his quarterly electricity bills for the last three years.

Year 1				Year 2				Year 3			
W	Sp	Su	A	W	Sp	Su	A	W	Sp	Su	A
76	60	36	64	84	68	44	72	92	76	52	80

He plots the figures as the blue line on the graph shown below.

Look at the figures. You can see that there is **seasonal variation** in them.

In which quarter does Dennis use the most electricity?

In which quarter does he use the least?

Why is this?

Dennis draws another line on the graph in red. This shows the **moving average**. Here is how he worked out the moving average.

	Year 1				Year 2				Year 3			
	W	Sp	Su	A	W	Sp	Su	A	W	Sp	Su	A
Moving average		59		61		63		65		67		69		71		73		75		

Notice that the first moving average is placed in the middle of the year, i.e. mid-way between spring and summer.

Where are the moving averages plotted on the graph?

Why is it done like this?

Dennis has averaged his values four at a time so this is called a **4-point** moving average.

Why is the red line much smoother than the blue one?

The red line is called the **trend line**. In this case the trend is increasing. A trend need not be a straight line but it can usually be described as rising, remaining steady, or falling.

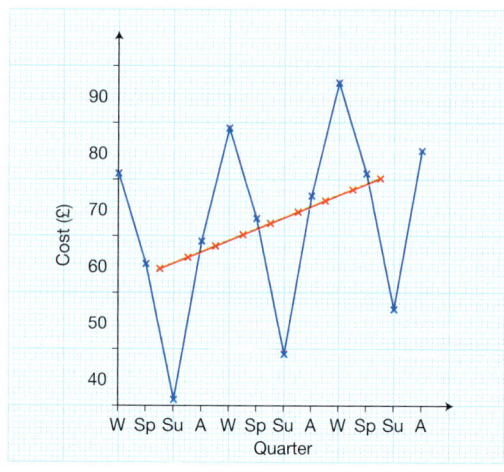

7: Data handling

1 a) How many values would you average at a time to find a 3-point moving average?

b) Find the 3-point moving averages for the following figures.

7 5 3 6 4 4 2 7 2 9 6 9 1 5 3 8 2 4

2 How many points would you use in your average

a) for monthly orders in writing an annual report?

b) for daily data when compiling a week-by-week comparison?

c) for a shopkeeper's daily takings if he opens six days each week?

3 Dennis wants to forecast the amounts of his next two bills.

a) State the value expected for the next moving average.

b) Use this value to predict the bill for winter of Year 4.

c) Repeat a) and b) to predict the bill for spring of Year 4.

If you draw a graph you can extend the trend line to find the next moving average.

4 The table shows the sales of ice-creams in thousands.

Year 1				Year 2				Year 3			
W	Sp	Su	A	W	Sp	Su	A	W	Sp	Su	A
1.4	3.2	5.8	2.2	1.8	3.6	6.2	2.6	2.2	4.0	6.6	3.0

a) Draw a graph to show the data.

b) Calculate the 4-point moving averages.

c) Draw a graph of the moving averages on the same axes.

d) Comment on the general trend.

e) Estimate the expected sales for winter and spring of Year 4.

5 The table shows absences from school for one class over 4 years.

Year 1			Year 2			Year 3			Year 4		
A	Sp	Su	A	Sp	Su	A	Sp	Su	A	Sp	Su
20	50	24	25	60	22	27	90	26	30	50	28

a) Draw a graph to show the data.

b) Calculate the 3-point moving averages.

c) Draw a graph of the moving averages on the same axes.

d) How many moving averages are affected by the '90' in spring of Year 3?

e) Suggest a reason for high absence in the spring term.

A spreadsheet program is ideal for calculating moving averages. Set up a spreadsheet to do the calculations that Dennis did.

7: Data handling

Bivariate data

Fatima has weighed and measured the people in her maths group, and drawn up this table.

	Angela	Brian	Colin	David	Elsa	Fatima	Gary	Helen
Weight (in stones)	7	10	13	14	10	11	15	9
Height (in inches)	65	69	72	71	64	66	76	60

The height and weight are called **variables**

Data like these are called **bivariate** ('bi' for 'two')

Fatima suspects that height and weight are related, so she draws this **scatter diagram**.

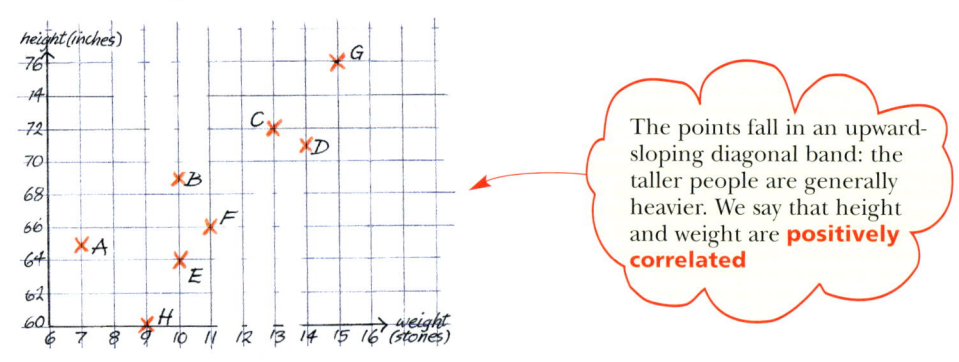

The points fall in an upward-sloping diagonal band: the taller people are generally heavier. We say that height and weight are **positively correlated**

- This scatter diagram shows a strong correlation because the points are close to a line. If the points are not clustered in a band there is little or no correlation.

- The correlation is positive in this case because the slope is upwards.

- A downward slope shows negative correlation.

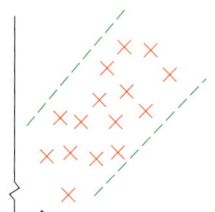

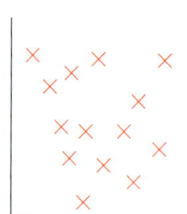

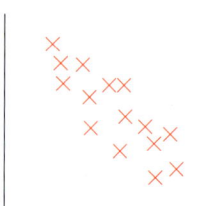

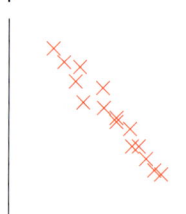

Which of these scatter diagrams shows

a) strong negative correlation?

b) moderate positive correlation?

c) little or no correlation?

No correlation implies no linear relationship – there may be some other relationship

What data would you expect to show each of these types of correlation?

7: Data handling

Exercise

1 A hospital keeps a records of the birth-weight and length of each baby born there. Here are the records for one day.

Name	Sex	Weight in kg	Length in cm	Name	Sex	Weight in kg	Length in cm
Faulkner	Boy	3.64	55	Wolf	Boy	4.18	63.5
Jones	Girl	3.49	53	Holt	Boy	2.7	59
Patel	Girl	4.46	54.6	Deane	Girl	3.47	55.9
Berry	Boy	3.84	53.4	Worth	Boy	2.98	53.3
Haynes	Boy	3.38	50.8	Murphy	Girl	3.81	61
Adams	Boy	3.69	58.4	McPhee	Boy	1.96	45.7
Gardner	Boy	2.61	47.5	Hughes	Girl	2.89	52
Roper	Girl	3.13	55.9	la Rue	Girl	3.98	57.2
Holmes	Boy	3.55	50.8	Marsh	Girl	2.55	49.2
McGrace	Boy	4.33	58.4	Mehta	Girl	2.6	50

a) Draw a scatter diagram and comment on the correlation.
b) Mark girls and boys in different colours. Does a pattern emerge?

2 The table shows the amount of vitamin B present in 100 g samples of spinach from crops treated with different amounts of fertiliser.

Fertiliser applied (kg per ha)	0	30	60	90	120	150	180	210
Vitamin B content (mg)	0.6	0.98	1.49	1.95	1.9	1.83	1.6	1.51

Draw a scatter diagram and comment on the correlation.

Draw a large scatter diagram of heights and weights to go on the wall. Keep measuring instruments handy and invite people to add themselves to the chart. (Remember that some people may be sensitive about their size, so do not press anyone who is unwilling.)

Height and weight charts are widely used by health workers. Try to borrow one, or just have a good look at it. Make sure you can see what it shows. Does it give any more information than your own wall chart?

Convert all the height and weight data in Fatima's table (opposite) into kilograms and centimetres.

Trace the scatter diagram, but label your scales in kilograms and centimetres. Read off the heights and weights of one or two people and check them against your table.

7: Data handling

Line of best fit

If the points on a scatter diagram lie in a narrow band there is a strong correlation. The stronger the correlation, the narrower the band. Provided there is some correlation, you can add a line of best fit to your diagram as in the example below.

Adding a line of best fit to your scatter diagram helps you to estimate the value of one variable, given the value of the other.

This table shows the mean heights of parents and the mean heights of their adult offspring.

Surname	Mean height of parents (cm)	Mean height of adult offspring (cm)
Wilson	161	163
Allan	164	167
Gupta	166	168
Smith	169	170
Lipton	174	173
Morris	179	177
Spicer	183	179
Gibson	183	183

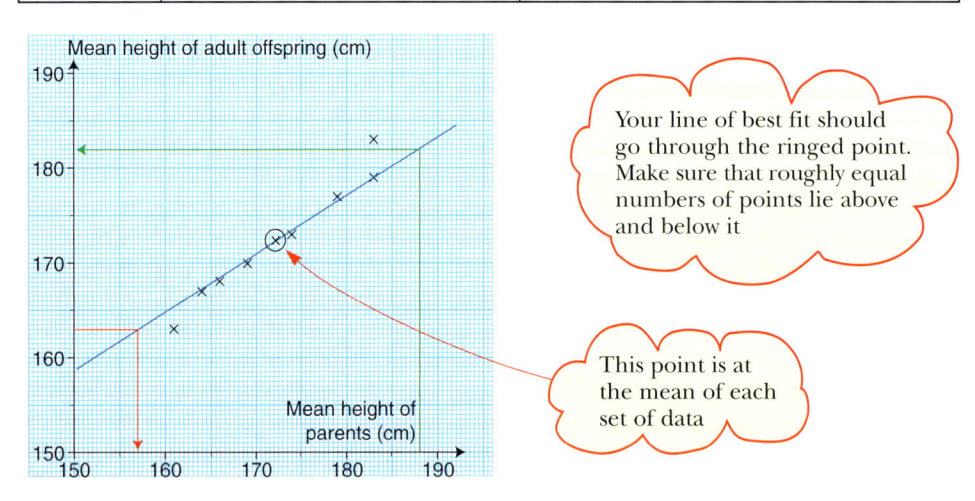

Your line of best fit should go through the ringed point. Make sure that roughly equal numbers of points lie above and below it

This point is at the mean of each set of data

? What does the diagram tell you?

? A group of brothers and sisters found that their mean height was 163 cm. Use the scatter diagram to estimate the mean height of their parents. (The red line will help.)

? A couple whose mean height is 188 cm are worried their children will be giants. They draw the green line on the diagram.

Do you think their children will all be 182 cm tall?

7: Data handling

Exercise

1 This table records the force P newtons required to move an object of mass M kg on a rough surface.

M	1	2	3	4	5	6	7	8
P	7.5	14.4	22.5	30.5	37.4	45.4	53.5	61.4

a) Draw a scatter diagram and add the line of best fit.
b) Estimate the force required to move a mass of 2.6 kg.
c) Estimate the mass you could move with a force of 40 newtons.

2 The people in Gil's chemistry class have done an experiment in which an acid reacts with a solid. They recorded the volume of gas given off during the reaction, and the loss in mass of the reactants. Here are their results.

Name	Alan	Bridie	Callie	Donna	Ed	Flick	Gil	Helen
Loss in mass (g)	0.060	0.032	0.120	0.083	0.090	0.160	0.140	0.107
Volume of gas (cm³)	45	24	90	62	35	120	105	80

a) Draw a scatter diagram of their results.
b) Comment on what your diagram shows.
c) One student made a mistake during the measuring. Who do you think it was?
d) Ignoring the data from the person who made a mistake, draw a line of best fit (by eye) for the data.
e) Use your line to estimate the mass of 100 cm^3 of the gas.

3 The table shows the number of hours of sunshine at a seaside resort one week in June, together with the number of people visiting the aquarium there.

Day	Hours of sunshine	No. of visitors
Mon	2.5	90
Tue	5.5	70
Wed	16	14
Thurs	13	28
Fri	10	44
Sat	0	96
Sun	7.5	64

a) Draw a scatter diagram.
b) Work out the mean of the number of hours' sunshine and the mean of the number of visitors. Plot the mean point on your diagram. Draw a line of best fit passing through this point.
c) How many visitors can the aquarium expect on a day when about 8 hours of sunshine are forecast?

Sometimes it does not make sense to draw a line of best fit. Draw two scatter diagrams for which this is the case.

7: Data handling

Finishing off

Now that you have finished this chapter you should be able to

★ draw and interpret pie charts and stem-and-leaf diagrams

★ calculate moving averages

★ interpret bivariate data by drawing scatter diagrams

★ recognise correlation when you see it and understand that it can be positive or negative, weak or strong

★ draw (by eye) and use lines of best fit.

Use the questions in the next exercise to check that you understand everything.

Mixed exercise

1 Salma, Bob and Carole start a new business. Salma contributes £30 000, Bob contributes £40 000 and Carole contributes £50 000.

a) Draw a pie chart for their wall to show what proportion each owns of the company.

The first year they make a profit of £18 000 which is to be shared in the same proportion as their contributions.

b) Use your pie chart to find how much profit each receives.

One year Carole receives £30 000 profit share.

c) How many pounds are represented by 1° on the pie chart?

d) How much does Salma receive that year?

2 Anita works in a small library. One week the library lent out 540 books in 4 categories. Anita drew a pie chart to show this information. The 'Thrillers' sector had an angle of 150°, the 'Biographies' sector had an angle of 70°, 45 'Reference' books were on loan and the other category was 'Romance'.

a) Calculate the number of 'Thrillers' on loan.

b) Find the angle of the 'Romance' sector.

3 Doug's grandfather recorded the distances he travelled by car each week for 22 weeks. The distances, to the nearest mile, are given below.

67 52 54 83 79 45 55 86 74 33 56
65 84 53 39 58 69 46 43 59 54 65

a) Draw an ordered stem-and-leaf diagram to display these data.

b) What was the shortest distance travelled?

c) How often did he travel more than 50 but less than 60 miles?

d) What was the longest distance travelled?

7: Data handling

Mixed exercise

4 Two firms make car batteries and a consumers' association tests 20 batteries from each firm. The battery lives in completed months are:

Longlife Batteries

45 49 47 55 58 40 46 50 51 54
55 61 63 68 51 44 49 58 57 50

Quickstart Batteries

65 45 50 58 60 65 47 68 55 58
62 69 48 62 56 50 58 46 64 66

a) Construct a sorted back-to-back stem-and-leaf diagram to show the data.

b) State the longest and shortest battery life of a 'Longlife' battery.

c) State the longest and shortest battery life of 'Quickstart' battery.

d) Compare the two sets of data and comment on whether there is a noticeable difference between the two brands.

5 The table shows the sales of Allan Inc.

Year 1				Year 2				Year 3			
W	Sp	Su	A	W	Sp	Su	A	W	Sp	Su	A
400	320	340	460	410	330	350	470	420	340	360	480

a) Draw a graph to illustrate the data.

b) Calculate the 4-point moving averages.

c) Draw a graph of the moving averages on the same axes.

d) Comment on the general trend.

e) Estimate the expected sales for winter and spring of the 4th year.

6 This table shows the height in metres above sea level and the temperature in Celsius, on one day at 9 different places in Europe.

Height (m)	1400	400	280	790	370	590	540	1250	680
Temperature (°C)	6	15	16	10	14	14	13	7	13

a) What is the mean height?

b) What is the mean temperature?

c) Plot a scatter diagram and describe the correlation.

d) Add a line of best fit.

e) Use your diagram to estimate the temperature at a height of 500 m.

f) Use your diagram to estimate the height of a place with a temperature of 8 °C.

Collect 5 articles from newspapers and magazines which present data. Do you think they are fair or misleading? Comment on each article. Look out for

- pictograms using larger symbols (rather than more symbols) to represent bigger numbers;
- graphs in which the vertical scales do not start at zero;
- misleading wording.

Eight

Graphs

Before you start this chapter you should

★ be able to substitute values into a formula
★ know how to use co-ordinates.

Look at the points on this straight line graph.

A is (1, 0): $x = 1, y = 0$
B is (2, 1): $x = 2, y = 1$
C is (3, 2): $x = 3, y = 2$

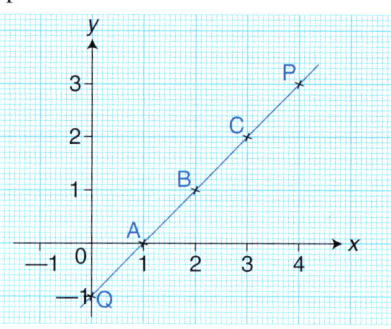

You can see that there is a pattern. In each case the value of y is 1 less than the value of x

$y = x - 1$

*This is called the **equation of the graph**.*

? *Do the points P and Q fit the same pattern?*

You often know the equation of a graph and want to draw it. In that case you start by making out a table of values, as in this example.

Example

Draw the graph of $y = 2x - 4$ for values of x between 0 and 4.

Solution

x	0	1	2	3	4
2x	0	2	4	6	8
-4	-4	-4	-4	-4	-4
y	-4	-2	0	2	4

2x is twice x

These are the values of x

−4 stays the same whatever the value of x

These are the values of y = 2x − 4

When x = 0, y = −4

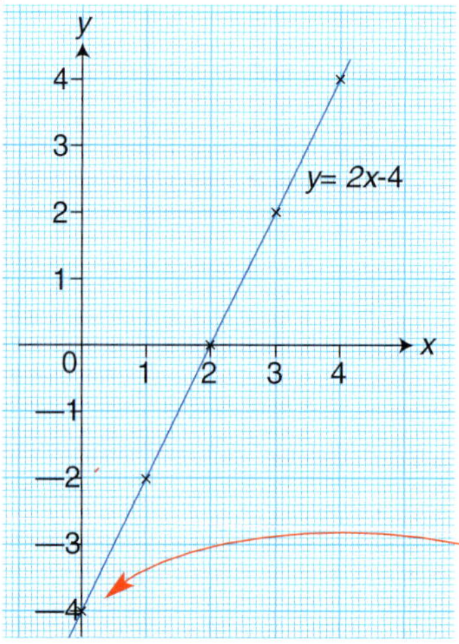

The points are plotted as crosses and then joined. In this case they lie on a straight line

In this graph the scales for x and y are the same. You can have different scales on the two axes.

*The line crosses the y axis at −4. This is called the **y intercept** or just the **intercept***

? Work out the values of y when $x = 1\frac{1}{2}$ and $3\frac{1}{4}$. Do these points also lie on the line?

8: Graphs

1 Look at these three sets of points.

A (–3, 0)   (–2, 1)   (–1, 2)   (0, 3)   (1, 4)
B (–2, –4) (–1, –3) (0, –2)   (1, –1) (2, 0)   (3, 1)
C (–2, –6) (–1, –3) (0, 0)   (1, 3)   (2, 6)

For each set of points,

a) choose x and y scales so that they fit nicely onto a piece of graph paper;
b) plot the points on a graph and join them with a straight line;
c) describe in words a formula to find y for a given value of x;
d) write an equation for each line, of the form $y = ...$;
e) write down the co-ordinates of two other points on the line and check that they satisfy the equation of the line.

2 a) Complete this table to find the co-ordinates of points on the straight line $y = 2x + 1$.

x	–2	–1	0	1	2	3
y	–3	–1				

b) Choose suitable scales, plot the points and join them with a straight line.
c) As you read along the table, x increases by 1 each time.
   Describe what happens to y as x increases.

3 A holiday in Spain costs £150, plus £20 per day at the Hotel Alhambra.

a) Make out a table showing the values of H when $D = 0, 5, 10, 15$.
b) Choose suitable scales and draw a graph showing the values of D along the x axis and H up the y axis.
c) Use your graph to find the cost of
   (i) a 7-day holiday and
   (ii) the number of days' holiday that you can get for £330.
d) Write down a formula for the total cost £H of a holiday with D days at the Hotel Alhambra.

The graph of the cost of sending a parcel against its weight looks like some steps.

Get a copy of the present postal rates and draw the graphs for 1st and 2nd class letters on the same sheet of paper.

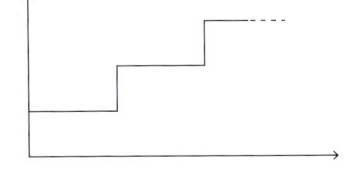

8: Graphs

Gradients and intercepts

Look at Hiral's table of values for $y = 4x - 8$.

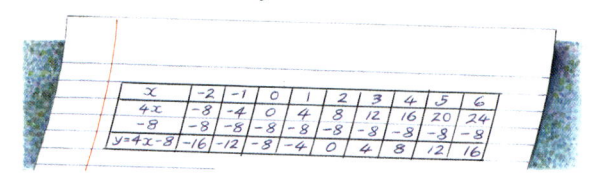

Hiral draws the graph using the same scales for his x and y axes. He finds that it looks very squashed because the line is steep. The steepness of a line is called its **gradient**.

Gradient = $\dfrac{\text{increase in } y}{\text{increase in } x}$

In this case,

y increases by 32 units, from -16 to $+16$;

x increases by 8 units, from -2 to $+6$.

So the gradient is $\dfrac{+32}{+8} = +4$.

Look at the graph and at the equation of the line,

$y = 4x - 8$

The gradient is 4

It crosses the y axis at -8. This is the intercept

? What are the gradient and intercept of the line $y = mx + c$?

These diagrams show you
- a horizontal line with gradient 0
- a line which slopes down from left to right, with negative gradient.

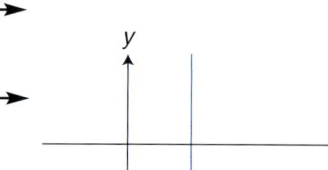

? What is the gradient of this line?

? What can you say about the gradients of a) parallel and b) perpendicular lines? The investigation on page 83 will help you.

Hiral decides to make the graph look less squashed. He uses different scales for the x and y axes.

It is often helpful to use different scales like this, but you must keep the same scale all the way along the x axis and all the way along the y axis, for both positive and negative values.

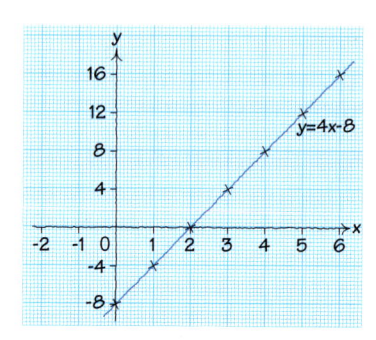

? Check that the gradient is still 4.

8: Graphs

1 The equation of a straight line is $y = 3x - 4$.

a) Make a table of values taking x from -2 to 4.

b) Choose suitable scales to draw the graph of the line.

c) Draw the graph of the line.

d) Calculate the gradient of the line.

e) What is the intercept of the line?

2 The equation of a straight line is $y = \frac{1}{4}x + 1$.

a) Make a table of values taking x from -2 to 4.

b) Choose suitable scales to draw the graph of the line.

c) Draw the graph of the line.

d) Calculate the gradient of the line.

e) What is the intercept of the line?

3 Find the gradient for each of these lines stating whether it is positive or negative. Write down the intercept. Then write down the equation of the line.

a)

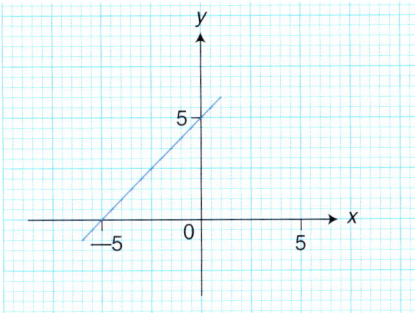

b)

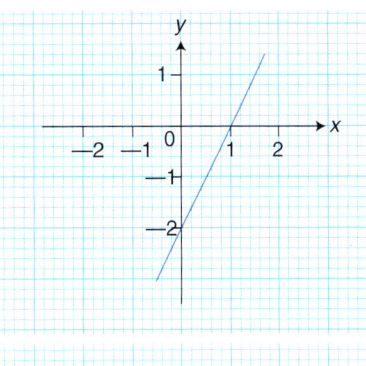

c)

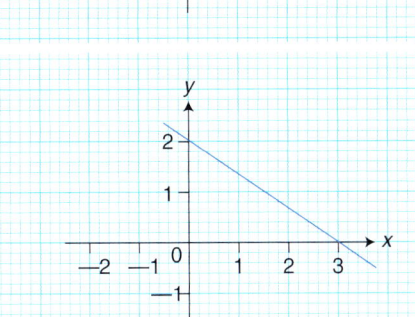

d)

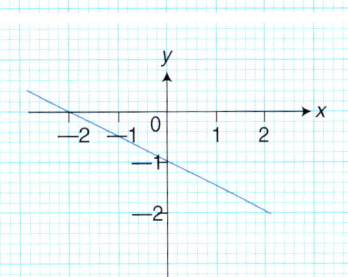

Investigation

On graph paper draw x and y axes from -5 to 5. Use the same scale for x and y.

Draw the graphs of

$y = 2x$, $y = 2x + 2$, $y = 3x$, $y = 3x - 2$, $y = -\frac{1}{2}x$, $y = -\frac{1}{3}$.

Which lines are parallel?

Which lines are perpendicular?

What can you say about the gradients of a) parallel and b) perpendicular lines?

8: Graphs

Obtaining information

The graph shows how much it costs to be a member of the OK Fitness Club. There is a joining fee and then an additional monthly charge.

? *Look at the graph. Can you tell what the joining fee is?*

Can you work out the monthly charge?

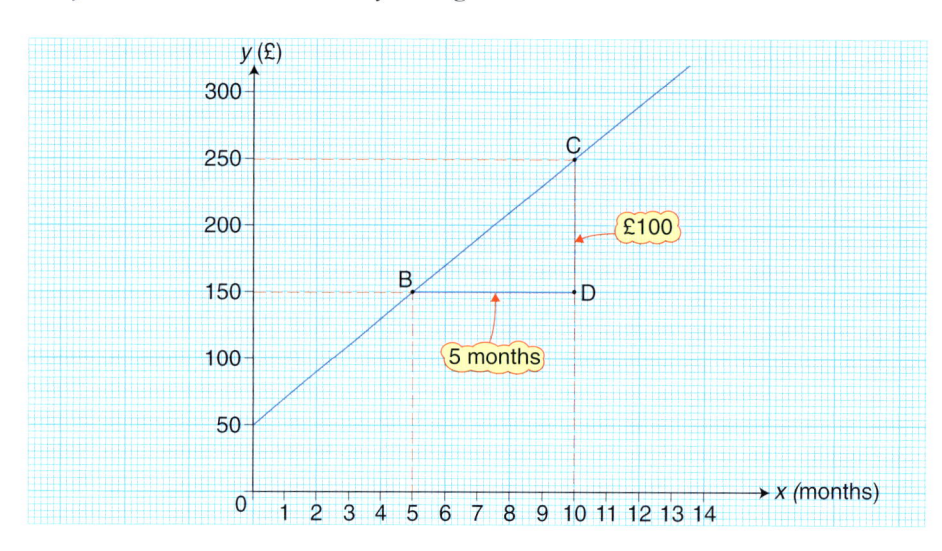

The joining fee is shown by the point where $x = 0$. This is called the **y intercept**. It is the price you pay right at the start.

Ali works out the monthly charge like this.

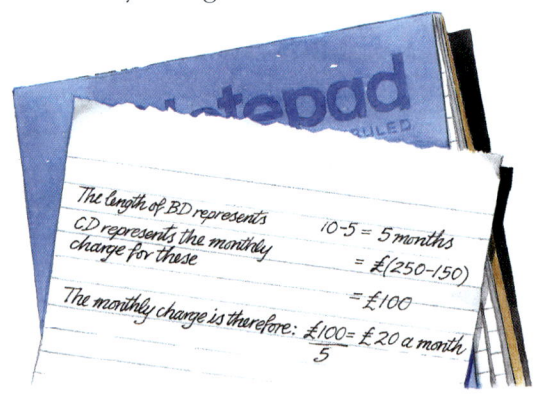

> To find pounds per month you divide pounds by months.

The length of BD represents $10 - 5 = 5$ months
CD represents the monthly charge for these $= £(250 - 150)$
$= £100$
The monthly charge is therefore: $\frac{£100}{5} = £20$ a month

? *What does the gradient of the graph represent?*

8: Graphs

1 The graph shows the price of taxi fares in a city. There is a fixed charge plus a rate per mile travelled.

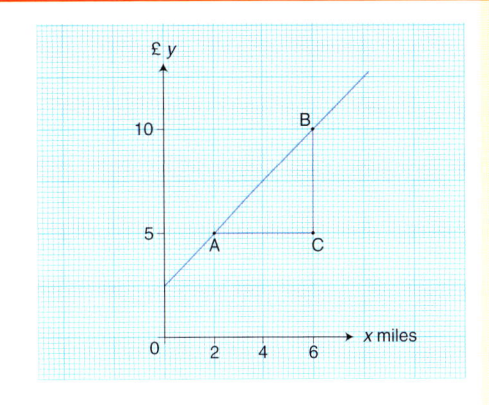

   a) What does each small unit on the *y* axis represent?
   b) What is the fixed charge?
   c) Write down the co-ordinates of A, B and C.
   d) Find the lengths of AC (in miles) and BC (in pounds).
   e) Use your answers to d) to find the rate per mile.

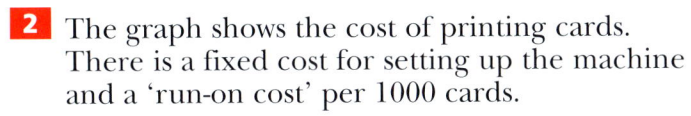

2 The graph shows the cost of printing cards. There is a fixed cost for setting up the machine and a 'run-on cost' per 1000 cards.

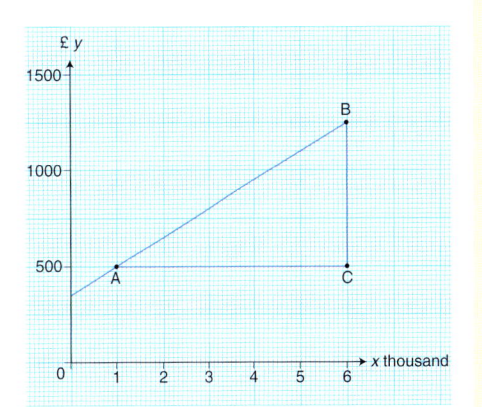

   a) What does each small unit on the *y* axis represent?
   b) What is the fixed setting-up charge?
   c) Write down the co-ordinates of A and B and so find the lengths of AC (in thousand cards) and BC (in pounds).
   d) Use your answers to c) to find the run-on cost per 1000 cards.
   e) Work out the cost of printing 10 000 cards.

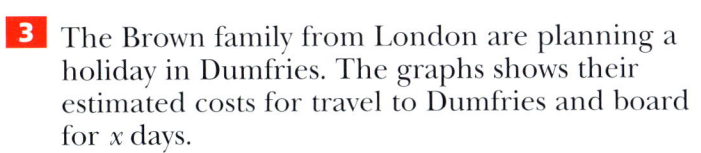

3 The Brown family from London are planning a holiday in Dumfries. The graphs shows their estimated costs for travel to Dumfries and board for *x* days.

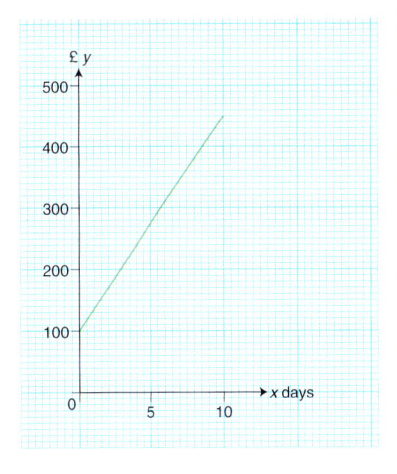

   a) How much do they estimate for travel?
   b) Find the gradient of the graph. What did they estimate for daily board?
   c) Write down an equation for the total cost £*C* for a holiday of *x* days.
   d) Use your equation to find the cost for 20 days.

A plumber, Alf, charges a call-out fee of £40 plus £12 per hour on the job. Write this as a formula for the cost £*C* of a job lasting *h* hours.

Another plumber, Beatrix, has a call-out fee of £32 and charges £16 per hour. Write this as a formula for the cost £*C*.

On the same piece of paper draw the graphs of *C* and *h* for the two plumbers. Which plumber is more expensive?

8: Graphs

Travel graphs

Catherine and her granny both live near a motorway, but at opposite ends of it. They agree to meet at a service station for lunch.

The red lines on the graph show Catherine's journey and the blue lines show Granny's.

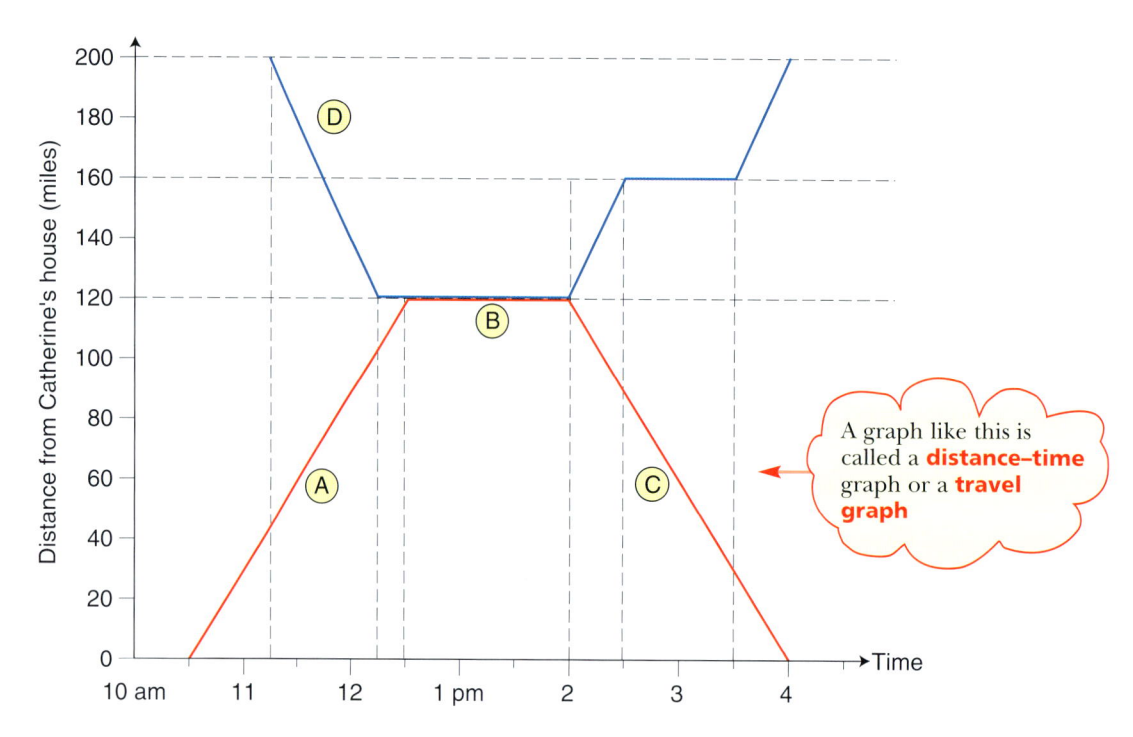

A graph like this is called a **distance–time** graph or a **travel graph**

? What is happening during part B of the graph?

Part A represents Catherine's journey to the service station.

She travels 120 miles in 2 hours.

The gradient of the line is $\dfrac{120 \text{ miles}}{2 \text{ hours}}$ = 60 miles per hour, and this is her speed.

Notice that the units $\dfrac{\text{miles}}{\text{hours}}$ can be written as miles/hour, miles hour^{-1} or miles per hour (mph)

The gradient of a distance–time graph gives the speed.

? What is Granny's speed on the way to the service station?

What is the meaning of a negative gradient on this graph?

? The lines on this graph are all straight.

What does this mean?

Is it realistic?

8: Graphs

Exercise

1 Use the graph on the opposite page to answer these questions.

a) When did Catherine arrive at the service station and how long did she stay?

b) How long did Granny have to wait for Catherine?

c) How long did Granny take to get home? What happened on the way?

d) Does Catherine drive more quickly, more slowly or at the same speed on the way home?

2 Karl leaves home at 12 noon for a cycle ride. The graph shows his journey.

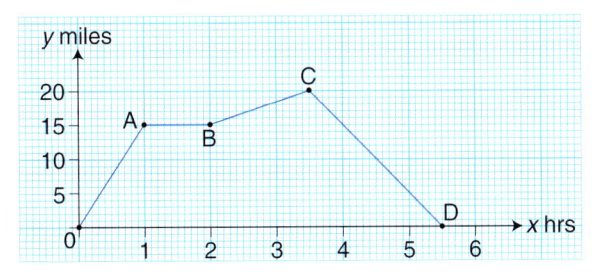

a) At what time is Karl furthest from home, and how far away is that?

b) Use the gradient to find his average speed in the part OA.

c) Describe what Karl might be doing from A to C.

d) The part CD has a negative gradient. Work out this gradient.

What does this tell you about this part of Karl's journey?

e) Karl's sister leaves home at 3.30 pm and drives at 40 mph along the same road.

Copy Karl's graph starting at 2.00 pm and add a line for his sister's journey. Where and when do they meet?

3 A train from Oxford to London stops at Didcot and Reading. Here is its timetable.

Oxford	Didcot		Reading		London
dep.	arr.	dep.	arr.	dep.	arr.
1200	1212	1215	1240	1245	1315

The first leg of the journey (Oxford to Didcot) is 10 miles. The second leg is 20 miles and the third is 35 miles.

a) Draw a travel graph of the journey using 1 cm for 5 minutes and 1 cm for 5 miles.

b) Find the average speed of the train for each part of the journey.

c) A non-stop London train travelling at 100 mph passes Oxford at 1230. Draw a line on your graph for this train. Find where and when it passes the first train.

Speed can be measured in miles per hour, kilometres per hour or metres per second (etc.).

Work out how to convert between these three units.

Give typical speeds of ten different objects (e.g. train, bicycle, snail) in all three units.

8: Graphs

Curved graphs

The graphs on the previous pages have all been straight lines. Sometimes you need to draw curves, as in the two examples on this page.

Example

Draw the curve $y = x^2 - 4x + 3$ for values of x between 0 and 4.

Solution

x	0	1	2	3	4
x^2	0	1	4	9	16
$-4x$	0	-4	-8	-12	-16
$+3$	+3	+3	+3	+3	+3
y	+3	0	-1	0	+3

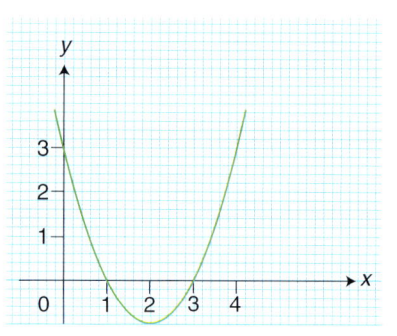

Why is it wrong to join the points with straight lines?

What are the values of x for which $y = 2$?

Example

Draw the curve $y = \dfrac{12}{x}$ for values of x from 1 to 6.

Solution

x	1	2	3	4	5	6
$y = \dfrac{12}{x}$	12	6	4	3	2.4	2

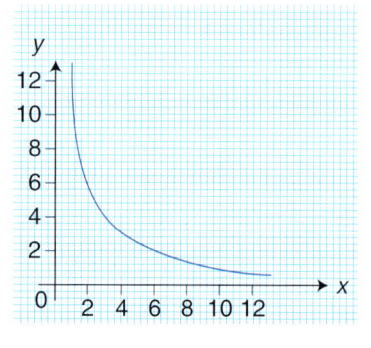

In this graph x cm and y cm could be the length and width of a rectangle of area 12 cm^2.

What is the length of the rectangle if the width is 2.5 cm?

What other meanings could x and y have?

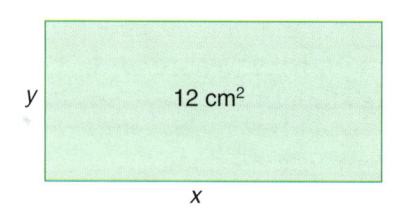

What happens to $y = \dfrac{12}{x}$ when x is 0?

88

8: Graphs

1 Alka throws a tennis ball up in the air. Its height, h metres above the ground, at time t seconds is given by

$$h = 20t - 5t^2$$

a) Copy and complete this table of values of h.

t	0	1	2	3	4
$20t$					80
$-5t^2$	0	−5			−80
h					0

b) Choose suitable scales and draw the graph of h against t.

c) Use your graph to estimate the times at which the ball is 10 m above the ground. Why do you get two answers?

d) Why is it not sensible to take values of t greater than 4?

For $-5t^2$ work out t^2 first and then multiply by -5: when $t = 4$, $4^2 = 16$ and $-5 \times 16 = -80$

2 A curve has equation $y = -4 + 5x - x^2$.

a) Copy and complete this table of values.

x	0	1	2	3	4	5
−4	−4					−4
$+5x$	0					25
$-x^2$	0					−25
y	−4					−4

Notice that this is -5^2 not $(-5)^2$.

b) Choose suitable scales and draw the graph.
c) Estimate the greatest value of y (i.e. the highest point of the curve).
d) Estimate the values of x for which $y = 1$.

3 A curve has equation $y = x^3 - 2x$.

a) Copy and complete this table of values.

x	−2	−1	0	1	2
x^3	−8				8
$-2x$	+4				−4
y	−4				4

Notice that $(-2)^3$ is -8

b) Choose suitable scales and draw the graph.
c) Estimate the values of x where the curve crosses the x axis.
d) Describe the symmetry of the curve.

On graph paper draw x and y axes from −5 to 5. Use the same scale for x and y. Plot the points (5, 0), (4, 3) and (3, 4). Now plot the images of these points when they are rotated anticlockwise, centre the origin, through 90°, 180° and 270°. You should now have 12 points on your graph paper.

What curve do they lie on?

Join them up by hand to draw the curve as well as you can.

8: Graphs

Finishing off

Now that you have finished this chapter you should be able to

* construct a table of values and use it to draw a graph
* find the gradient and intercept of a straight line graph
* know when gradients are positive, zero and negative
* obtain information from graphs using co-ordinates of points
* find a fixed value of starting value from a graph
* draw and understand travel graphs.

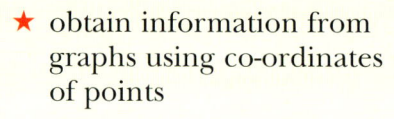

Mixed exercise

1 The following equations represent straight lines.

a) $y = x + 1$ b) $y = 2x - 3$ c) $y = \frac{1}{2}x + 2$ d) $x + y = 5$

For each one,

(i) construct a table of values from $x = -2$ to 4;

(ii) choose suitable scales;

(iii) draw the graph;

(iv) calculate the gradient of the line;

(v) state the intercept on the y axis.

2 A curve has equation $y = x^2 - 5$.
a) Copy and complete this table of values.

x	–3	–2	–1	0	1	2	3
x^2	+9		+1	0			
–5	–5	–5		–5			
y	+4			–5			

b) Draw the graph. Use scales of 2 cm for 1 unit on each axis.
c) Find the value of y when $x = 1.2$.
d) Estimate the values of x for which $y = 2$.

90

8: Graphs

Mixed exercise

3 The diagram shows a box. The base is a square of side x cm. The height is y cm. The volume of the box is 36 cm³.

  a) Explain why $x^2y = 36$.

  This can be rewritten as $y = 36/x^2$.

  b) Construct a table of values of y for values of x from 1 to 6.

  c) Draw the graph of $y = 36/x^2$.

  d) Use your graph to estimate the value of x for which $y = 5$.

4 The graph shows how the annual cost of running Mary's car depends on the number of miles she travels.

  a) What is the y intercept on the graph?

  b) What is this money spent on?

  c) Find the gradient of the graph and write it in pence per mile.

  d) What is this money spent on?

5 The graph shows the cost of driving lessons when they are paid for by the hour. The points A and B show special rates for 10 hours and 25 hours if you book in advance.

  a) Find the price for 10 hours at the normal hourly rate.

  b) Debbie pays for 25 hours in advance. How much does she save by paying this way?

  c) Find the normal price per hour.

  d) Debbie's brother thinks he will only need 20 hours, so he books 2 sets of 10 hours. He then finds he needs another 5 hours, and pays for them at the normal rate.

  How much has he paid for his 25 hours?

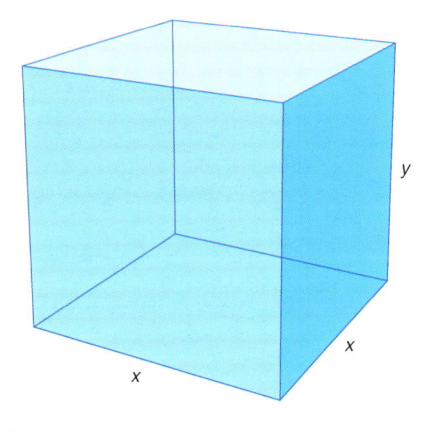

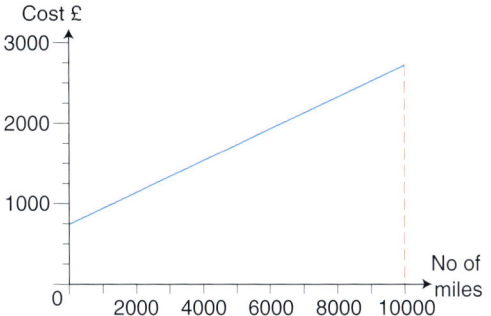

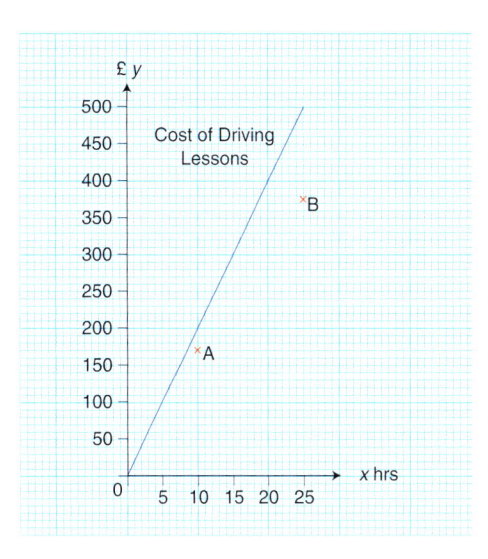

6 Ben throws a boomerang. After t seconds its distance from Ben is y metres, where $y = 25t - t^3$.

  a) Make a table for $t = 0$ to 5. Use separate rows for t, $25t$ and t^3 and subtract the third row from the second to find y.

  b) Draw a graph using 2 cm for 1 unit along the t axis and 2 cm for 10 units along the y axis.

  c) How long does it take the boomerang to return to Ben?

  d) What is its greatest distance from Ben?

Nine

Ratio and proportion

Before you start this chapter you should be able to

★ find equivalent ratios

★ write a ratio in its simplest form

★ divide a quantity in a given ratio.

Use the questions in the next exercise to check that you still remember these topics.

Revision exercise

1 In each of these, write down the missing number.

a) 1:2 is the same as 5:?
b) 25:1 is the same as 75:?
c) 1:10 is the same as ?:40
d) 5:100 is the same as 1:?
e) 2:3 is the same as 12:?
f) 80:5 is the same as ?:1
g) 3.5:7 is the same as 1:?
h) 4:200 is the same as 1:?
i) 7.5:22.5 is the same as 1:?
j) 2:6:10 is the same as 1:?:?

2 In each of these pairs of similar triangles, find the length of the side marked with a question mark. (Note: the triangles are not drawn to scale.)

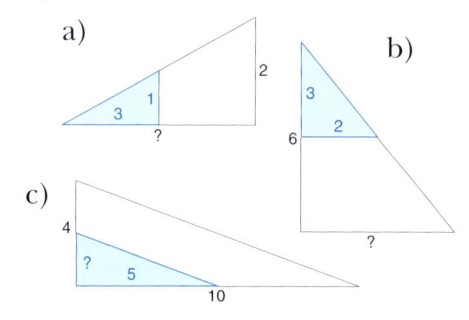

a)
b)
c)

3 Write each ratio in its simplest form.

a) 9:12
b) 20:5
c) 10:30
d) 250:50
e) 35:21
f) 25:40
g) 8:20
h) 60:40
i) 24:18
j) 75:175
k) 140:56
l) 4:6:10

4
a) Write 120:30 in the form ——:1.
b) How many minutes are there in 2 hours?
c) Write the ratio 2 hours : 30 minutes in its simplest form.

9: Ratio and proportion

Revision exercise

5 Write each ratio in its simplest form.
   a) 1 hour : 15 minutes   b) 50 grams : 1 kilogram
   c) 70 cl : 1 litre   d) £3 : 60 pence   e) 3 cm : 15 mm
   f) 2 cm : 1 km   g) 2.1 m : 1.4 m   h) $1\frac{1}{2}$ seconds : 6 seconds
   i) $12\frac{1}{2}$ hours : $7\frac{1}{2}$ hours

6 a) Write the ratio 15:20 in its simplest form.
   b) How many sheets can printer A print in 1 hour?
   c) How many sheets can printer B print in 1 hour?
   d) Write the number of sheets per hour as a ratio (printer A : printer B).
   e) Write this ratio in its simplest form. What do you notice?

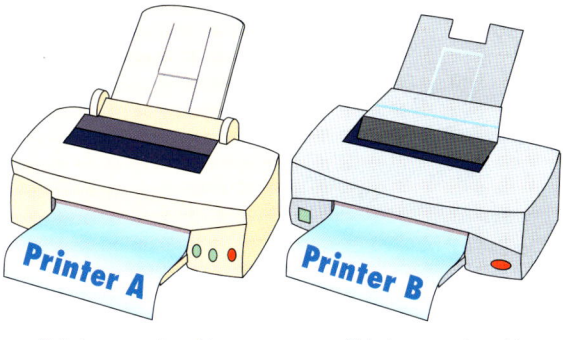

Printer A — Prints one sheet in 15 seconds
Printer B — Prints one sheet in 20 seconds

Mix 20 ml plant food with 1 litre of water

7 a) How much of this plant food would you mix with 3 litres of water?
   b) What is the correct mixing ratio of plant food to water?
   c) Write this ratio in its simplest form.

8 Carlos and Darren provide capital for a business in the ratio 2:3.
   a) What fraction of the capital does Carlos provide?
   b) What percentage of the capital does Darren provide?
   c) Darren provides £22 500. How much does Carlos provide?

9 a) Share £300 in the ratio 3:2.
   b) Share £210 in the ratio 2:1.
   c) Share £6000 in the ratio 5:3.
   d) Share £750 in the ratio 1:3.
   e) Share £585 in the ratio 4:5.
   f) Share £3.50 in ratio 5:2.

There are many different recipes for non-alcoholic fruit punch to serve at parties. The ingredients for one variation are shown below.

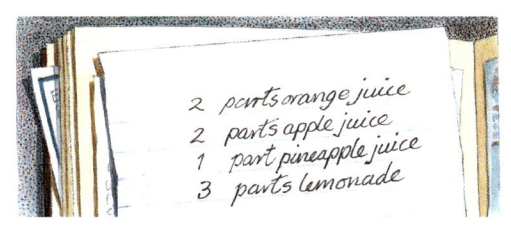

2 parts orange juice
2 parts apple juice
1 part pineapple juice
3 parts lemonade

Test the recipe and adjust the proportions to your taste.

Work out the approximate quantities (in litres) of each ingredient that you would need to serve your modified recipe to 30 people on a hot day.

9: Ratio and proportion

Using ratio

A new 24-hour satellite TV company, NFS, has sent out a marketing leaflet. It includes this pie chart showing the ratio of the various types of programmes that it will broadcast.

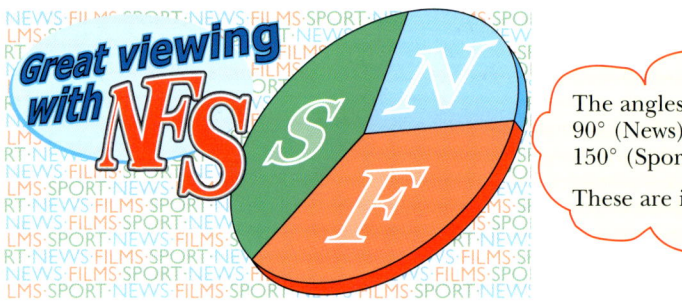

The angles on the chart are 90° (News), 120° (Film) and 150° (Sport).

These are in the ratio 3:4:5

Julian works out how many hours of each type of programme there should be in a week.

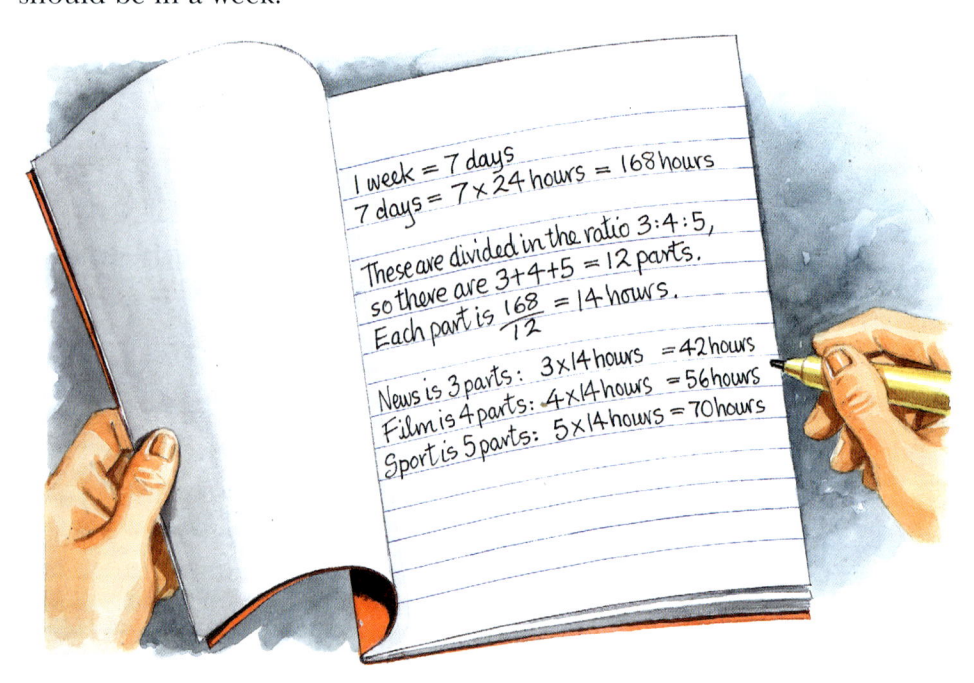

1 week = 7 days
7 days = 7 × 24 hours = 168 hours

These are divided in the ratio 3:4:5,
so there are 3+4+5 = 12 parts.
Each part is $\frac{168}{12}$ = 14 hours.

News is 3 parts: 3 × 14 hours = 42 hours
Film is 4 parts: 4 × 14 hours = 56 hours
Sport is 5 parts: 5 × 14 hours = 70 hours

Julian does a survey of the channel's programmes over 2 days, a Monday and a Tuesday. Here are his results.

Type of programme	News	Film	Sport
Number of hours	14	15	19

? Do these numbers match the ratio given in the pie chart?

? Does this mean that NFS is misleading its customers?

9: Ratio and proportion

1 In one week, 88 babies are born in a hospital.

A nurse draws this diagram to show the numbers of boys and girls.

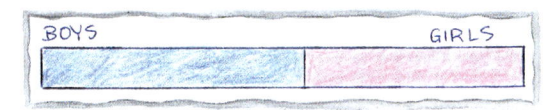

a) What is the ratio of boys to girls?
b) What fraction of the babies are boys?
c) How many boys are there?
d) What fraction of the babies are girls?
e) How many girls are there?

2 Charlotte and Thomas start a business. Charlotte provides £4500 and Thomas provides £7500.

a) Write this ratio in its simplest form.

Each month Charlotte and Thomas share profits in the ratio 3:5.

b) In May the profit is £400. How much does each get?
c) In June Charlotte gets £240. Work out the total profit.
d) In July Charlotte gets £156 less than Thomas. How much does each receive?

3 A factory produces recycled paper and prints this label.

This means that 40% of the material in the paper has been used before.

The rest of the paper is made from other material.

Recycled paper
Contains 40% post-consumer waste

Each skip contains 1 tonne of post-consumer waste.

a) The company uses one of these skips when making 1 batch of recycled paper.

What is the weight of paper in 1 batch?

b) What is the ratio of post-consumer waste to other material in the paper?

4 In a football competition the allocation of tickets between the home club and the away club is in the ratio 6:1.

a) For one match, there are 38 500 tickets. How many tickets does each club receive?
b) For another match, the home team gets 21 600 tickets. How many does the away team get?
c) The away team in the match in part b) returns half its tickets, which the home team then sells. In what ratio are the tickets sold?

Jules is a chef. He makes pastry by mixing flour, lard and margarine in the ratio 8:3:1 (before adding water).

Design and make a table to tell Jules the required amount of each ingredient when he is cooking for different numbers of people. (He uses 120 g lard for 6 people.)

9: Ratio and proportion

Unitary method

Jade needs some mineral water.

Her local shop sells these two sizes.

? *Which do you think is the better buy?*

Jade compares the prices by working out the price of 1 litre from each container.

The 2 litre container costs 60p:

$$\text{price of 1 litre} = \frac{60p}{2} = 30p$$

The 5 litre container costs 140p:

$$\text{price of 1 litre} = \frac{140p}{5} = 28p$$

The 5-litre container is the better buy: it has the lower price per litre.

Josh buys a computer printer for £170. This is the price after a 15% discount. Josh works out the price before discount like this:

$$85\% \text{ is } 170$$

$$1\% \text{ is } \frac{170}{85}$$

$$100\% \text{ is } \frac{170}{85} \times 100 = 200$$

The usual price is £200.

? This method, of first finding 1%, is called the **unitary method**.

Why is this a useful technique?

Example

200 g of meat costs £1.44. How much does 340 g cost?

Solution

200 g costs £1.44

1 g costs $\frac{£1.44}{200}$

340 g costs $\frac{£1.44}{200} \times 340 = £2.448$

This has to be rounded to give a whole number of pence.

340 g of meat costs £2.45 to the nearest penny.

9: Ratio and proportion

1 Work out which is the better/best buy.

a) Cola: £1·84 (4-pack) / £2·69 (8-pack)

b) £0·96 / £1·65

c) Bran: 250g £0·69 / 500g £1·29 / 750g £1·99

2 The mass of 3 cm^3 of gold is 57.9 g. Work out the mass of

a) 5 cm^3 of gold   b) 30 cm^3 of gold.

3 Verity, Misha and Kay are all paid at the same hourly rate.

Verity earns £168 for a 35-hour week.

a) Work out the hourly rate of pay.
b) Misha works a 38-hour week. How much does he earn?
c) Kay earns £201.60 a week. How many hours does she work?

4 Ian can drive 153 miles on 18 litres of petrol.

a) How far can he drive on 30 litres of petrol?
b) How many litres of petrol does he need to drive 119 miles?
c) He has 25 litres of petrol in the tank. Will this be enough to drive to see his sister, who lives 110 miles away, and back?

5 Emma hires a barge for a week.

a) How much does it cost?
b) How many kilometres can she travel?

Paul has a budget of £190.

c) What is the greatest number of days he can have?
d) How many kilometres can he travel?

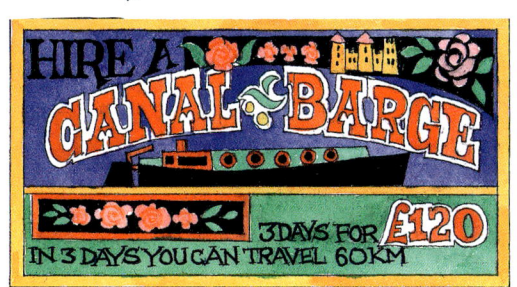

HIRE A CANAL BARGE — 3 DAYS FOR £120 — IN 3 DAYS YOU CAN TRAVEL 60 KM

6 Luke makes 12 buns. He uses 300 g of flour, 120 g of sugar, and 120 g of butter.

a) How much of each ingredient would he use to make 4 buns?
b) How much flour would he need for 20 buns?
c) What is the greatest number of buns he can make with 100 g of butter?

Go to a supermarket and find an item which you can buy in at least 3 sizes.

Is it the case that the larger the quantity the lower the unit cost?

9: Ratio and proportion

Changing money

How do you get foreign money?

Lisa is going on holiday to the USA. She wants to change £400 into dollars. She goes to this bureau de change.

Rates	£1=
Israel	5.95 shekels
S Africa	15.9 rand
Japan	185 yen
USA	$1.58

Each pound is worth $1.58, so £400 is worth

400 × 1.58 = 632

Lisa receives $632 for her £400.

How many dollars would Lisa get for a) £200? b) £750?

Note: The bureau de change will probably charge commission when changing Lisa's £400 into dollars. A typical rate is 2%, and 2% of £400 is £8. So Lisa would actually pay £408 for her $632.

Lisa arrives in the USA and buys a meal costing $22.

Each $1.58 is worth £1. Lisa works out the price in pounds by finding out how many times $1.58 goes into $22:

22 ÷ 1.58 = 13.92

> This is rounded to 2 decimal places so it is correct to the nearest penny

Lisa's meal costs £13.92.

Find the cost in pounds of a) a jacket costing $55, b) a coach ticket costing $96.

Lisa makes this conversion table to help her to convert dollars into pounds. She has worked out some of the conversions (to the nearest penny).

$1	£0.63	$15	
$2	£1.27	$20	
$3		$30	
$4		$40	
$5		$50	
$10	£6.33	$100	

Copy and complete this conversion table.

How can she use it to work out $90 in pounds?

9: Ratio and proportion

Use the exchange rates opposite for each of these questions.

1 Rea, Kate and Callum go to Japan.

a) How many yen does Rea get for £40?

b) How many yen does Callum get for £95?

c) How much does Kate pay for 18 500 yen?

d) A trip to Mount Fuji costs 8000 yen. How much is this in pounds and pence?

e) A dish of noodles costs 750 yen. How much is this in pounds and pence?

2 Paul is going to Israel. He changes £300 into shekels before he goes.

a) How many shekels does he get?

b) He pays 2% commission. How much does he pay in total?

c) Paul buys these items whilst in Israel.

185 shekels

650 shekels

What is the cost of each item in pounds and pence?

d) At the end of his holiday Paul changes 124 shekels back into pounds. He pays no commission. How much does he get?

3 Lucy goes to South Africa. She makes this conversion table.

a) Copy and complete the table.

b) Explain how she would use it to find the cost in pounds and pence of an item costing (i) 7 rand (ii) 45 rand.

c) Work out the cost of these items in pounds and pence.

Rand	£	Rand	£
1	£0.06	15	
2		20	
5		50	
10	£0.63	100	

Choose a country that you would like to visit.

Find out the name of its currency, and the current exchange rate against the pound.

Draw up a table to help you to convert prices quickly into pounds whilst you visit the country.

9: Ratio and proportion

Distance, speed and time

? *What is the speed limit on motorways?*

How fast does an aeroplane fly?

How many metres per second can you run?

Speed is usually measured in miles per hour (m.p.h.), kilometres per hour (km/h) or metres per second (m/s).

Tim drives at 100 km/h on the motorway.

? *Do you think that he drives at exactly* 100 km/h *all the time?*

Tim drives 90 km on country roads in $1\frac{1}{2}$ hours.

What is his average speed in km/h?

$$\text{Average speed} = \frac{\text{distance covered}}{\text{time taken}}$$

$$= \frac{90}{1\frac{1}{2}} = 60$$

Tim's average speed is 60 km/h.

? *Kanwal does the same journey in* $1\frac{1}{4}$ *hours.*

What is her average speed?

Tim's average speed on a motorway is 100 km/h.

How long does a motorway journey of 225 km take?

$$\text{Time taken} = \frac{\text{distance covered}}{\text{average speed}}$$

$$= \frac{225}{100} = 2.25 \text{ or } 2\frac{1}{4}$$

The time taken is $2\frac{1}{4}$ hours (or 2 hours 15 minutes).

> Be careful here! .25 hours is not 25 minutes. To change 0.25 hours into minutes you multiply by 60
> $0.25 \times 60 = 15$ so
> 0.25 hours = 15 minutes

? *How long does it take Tim to travel* 60 km*?*

How far does Tim travel in $1\frac{3}{4}$ hours?

$$\text{Distance covered} = \text{average speed} \times \text{time taken}$$

$$= 100 \times 1\frac{3}{4} = 175$$

Tim travels 175 km in $1\frac{3}{4}$ hours.

? *How far does Tim travel in 45 minutes?*

9: Ratio and proportion

1 What distance is covered by

   a) Linton cycling at 50 km/h for 2 hours?

   b) Jovanka driving at 70 km/h for $1\frac{1}{2}$ hours?

   c) Philip flying at 880 km/h for $2\frac{1}{4}$ hours?

   d) Liz running at 15 km/h for 1 hour 20 minutes?

2 Hana has four meetings today. This is her schedule.

Work out the average speed

   a) between London and Milton Keynes (84 km apart)

   b) between Milton Keynes and Leicester (81 km apart)

   c) between Leicester and Sheffield (105 km apart).

TUESDAY 6
MEETING
Leave London 10.00
Arrive Milton Keynes 11.30
MEETING
Leave Milton Keynes 12.30
Arrive Leicester 13.45
MEETING
Leave Leicester 14.45
Arrive Sheffield 16.00
MEETING

3 Work out the time it takes to

   a) cycle 75 km at 30 km/h

   b) fly 2250 km at 600 km/h

   c) drive 100 km at 60 km/h

   d) run a marathon (42.2 km) at 20 km/h.

4 Darren from Liverpool and Vicky from Hull drive to Manchester to meet for lunch.

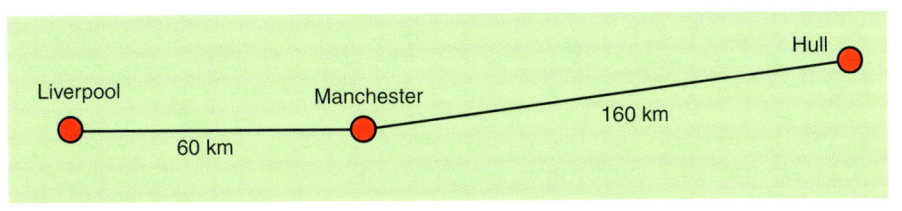

Liverpool — 60 km — Manchester — 160 km — Hull

   a) Darren leaves home at 1140 and his average speed is 60 km/h. What time does Darren arrive in Manchester?

   b) Vicky leaves at 1045 and expects to take 2 hours. What will her average speed be?

   c) Vicky's journey takes 30 minutes longer than planned. What is her average speed?

   d) How long did Darren have to wait on his own?

9: Ratio and proportion

Finishing off

Now that you have finished this chapter you should be able to

★ relate a ratio to fractions, decimals or percentages

★ write a ratio in its simplest form

★ solve simple problems using ratio

★ compare prices and work out the 'best buy'

★ change money

★ solve problems with distance, speed and time.

Use the questions in the next exercise to check that you understand everything.

Mixed exercise

1 Write each ratio in its simplest form.
 a) 5:20   b) 24:8
 c) 60:100   d) 12:20
 e) 105:35   f) 18:27
 g) 32:24:16   h) 35:84:91
 i) 2 hours : 10 minutes
 j) 500 ml : 1 litre
 k) $6\frac{1}{4}$ miles : 5 miles
 l) 5.4 kg : 3.6 kg

2 A shade of orange paint is made by mixing red and yellow in the ratio 1:3.
 a) What fraction of the orange paint is made up of red paint?
 b) What percentage of the orange paint is made up of yellow paint?
 c) How many tins of yellow would you mix with 2 tins of red?
 d) How many tins of red would you mix with 12 tins of yellow?

3 a) Lucy cycles for 45 minutes at an average speed of 30 km/h. How far does she travel?
 b) Anant is a airline pilot. He has 900 km to cover in $1\frac{1}{4}$ hours. What average speed does he need to attain?
 c) Tracey has 50 km to drive. She thinks her average speed will be 65 km/h. How long, to the nearest 5 minutes, will it take her?

4 Asif plots the distance his coach has travelled against time. This journey is broken into 3 stages by 2 breaks. This is his graph. 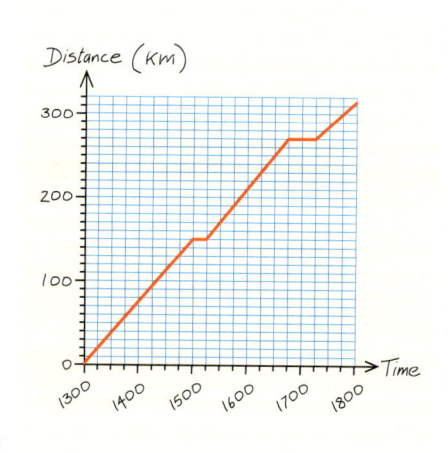
 a) What is the average speed for the first stage of the journey?
 b) How long was the first break?
 c) What is the average speed for the second stage of the journey?

The total distance covered is 315 km.
 d) What is the total time taken including breaks?
 e) What is the average speed for the whole journey?

9: Ratio and proportion

Mixed exercise

5 Brass is made by mixing copper and zinc in the ratio 7:3.
   a) How much brass can be made with 280 g of copper?
   b) How many grams of each metal are needed to make $\frac{1}{2}$ kg of brass?

6 Sue and Bob are going on holiday to the U.S.A. The exchange rate is £1 = $1.60.
   a) How many dollars will Sue get for £110?
   b) Bob buys $200 and pays 2% commission charge. Work out the cost in pounds and pence.
   c) Sue pays $30 for a coach ticket in New York. How much is this in pounds and pence?

7 Emily and Anna start up a babysitting agency. Emily works 3 nights and Anna works 5 nights. (They both work on Saturdays.)

They share the profits in the ratio of the nights they work. In the first month Anna gets £60 more than Emily.
   a) How much does each receive?
   b) What percentage of the profit does Emily receive?
   c) What percentage of the profit does Anna receive?
   d) After a year the profits have doubled.
      What percentage of the profit does Emily now receive?

8 Sanjay travels at 60 mph on a motorway.
   a) How long does it take him to travel 165 miles?
   b) How far does he travel in 1 hour and 20 minutes?

9 Martha wants to learn to drive. Which driving school offers her the best deal?

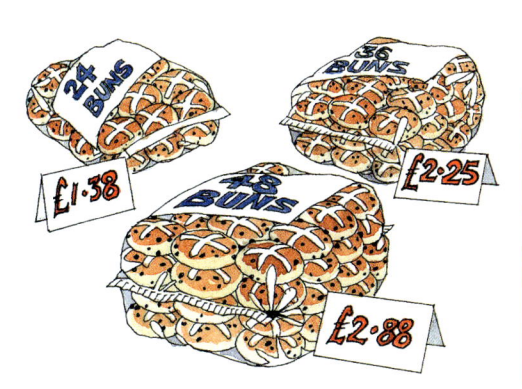

MOTOR MO Motoring School — 9 lessons £160 — 10th lesson FREE

Drive Dream — 6 lessons £99

Learn with Len — 4 lessons £69

10 Work out which is the best buy.

24 BUNS £1.38
36 BUNS £2.25
48 BUNS £2.88

A company's computer is used to analyse data records. Analysing 540 records takes the computer $\frac{1}{4}$ hour. How long does it take the computer to analyse 72 more records? (Assume each record requires the same amount of computer time.)

Ten

Grouped data

Before you start this chapter you should be able to

★ make a frequency table for a set of data

★ draw vertical line charts and bar charts

★ work out the mode, median, mean and range of a set of numbers.

Clare and Fulvio measure the height, h cm, of drama group members.

165.3	176.0	176.1	168.1	169.5
165.9	167.4	170.0	178.9	169.9
169.2	173.8	165.8	168.4	185.8

They want to display the data and Fulvio starts to draw a vertical line chart. He has not got very far when Clare stops him.

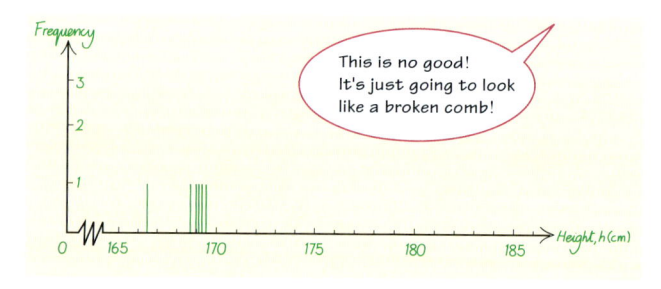

This is no good! It's just going to look like a broken comb!

? *Do you agree with Clare's comment?*

Clare draws up this grouped frequency table.

Height, h (cm)	$165 \leq h < 170$	$170 \leq h < 175$	$175 \leq h < 180$	$180 \leq h < 185$	$185 \leq h < 190$
Frequency (number of people)	9	2	3	0	1

Notice that Clare uses different symbols at each end of an interval.

$\leq$ means 'is less than or equal to'

$<$ means 'is less than'

? *Clare is 169.9 cm tall. Jem is 170.0 cm. Which groups are Clare and Jem in?*

Clare now draws a graph to illustrate the data. It is called a histogram.

Notice that

- there are no gaps between the bars
- the horizontal scale is continuous
- the vertical scale is described as 'frequency per 5 cm interval'. This means that bars of equal *area* represent equal frequencies.

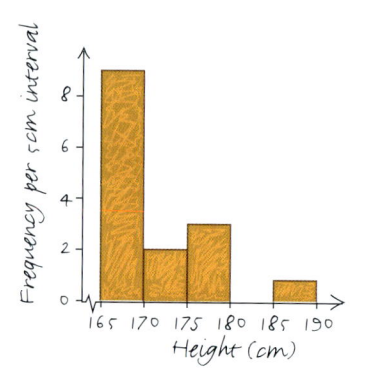

10: Grouped data

1 A company buys new cars for its fleet in batches of 12, and sells them after exactly one year. The mileage, m, on cars in one batch for sale are as follows.

23 147	18 219	46 351	33 337
16 517	23 474	25 012	29 830
29 216	9051	41 607	31 412

a) Make a grouped frequency table for these figures using intervals $0 \leq m < 10\,000$, $10\,000 \leq m < 20\,000$ and so on.

b) Draw a histogram to illustrate the data.

c) Two garages want to buy the cars. Here are the deals they offer.

Garage A

Mileage, m	Price (£)
$0 \leq m < 10\,000$	12 000
$10\,000 \leq m < 20\,000$	11 000
$20\,000 \leq m < 30\,000$	10 000
$30\,000 \leq m < 40\,000$	9 000
$40\,000 \leq m < 50\,000$	8 000

Garage B

Mileage, m	Price (£)
$0 \leq m < 20\,000$	13 000
$20\,000 \leq m < 40\,000$	10 000
$40\,000 \leq m < 60\,000$	7 000

Which garage should the company deal with if

(i) the deal is for all the cars?

(ii) the deal is for any number of cars?

State the total price in each case.

2 Car speedometers are being tested for accuracy at a road safety laboratory. In one 75 mph test, 30 speedometers are tested and their readings are as shown below.

71.8	73.6	72.8	74.7	73.7	74.8	74.1	75	74.3	75.2
73.8	72.1	76.7	73.9	74.8	73.2	76.8	76.2	74.7	74.4
72.3	76.9	72.9	75.4	74.9	77	74.1	75.1	74.2	77.2

a) Group these readings into classes 70 but less than 71 mph, 71 but less than 72 mph, etc. Think carefully about how you label these classes.

b) Draw a histogram to show these readings.

c) How many of the readings are in error by more than 4%?

10: Grouped data

Grouping rounded data

A bus company plans to start a new long distance service between two cities. They need to know how long the journey will take, so they do 15 trial runs. Each time the driver tells the company how long he took, to the nearest minute.

165	176	176	168	170
166	167	170	179	170
169	174	166	168	186

165 minutes 20 seconds is recorded as 165. 165 minutes 50 seconds is recorded as 166

Back in the office Tom has to display the data. He starts by making this grouped frequency table.

TIME (MINUTES)	165 – 169	170 – 174	175 – 179	180 – 184	185 – 180
FREQUENCY (NO. OF BUSES)	7	4	3	0	1

When Tom draws the histogram he marks the ends of the bars at $164\frac{1}{2}$, $169\frac{1}{2}$, $174\frac{1}{2}$ and so on. The diagram below shows why.

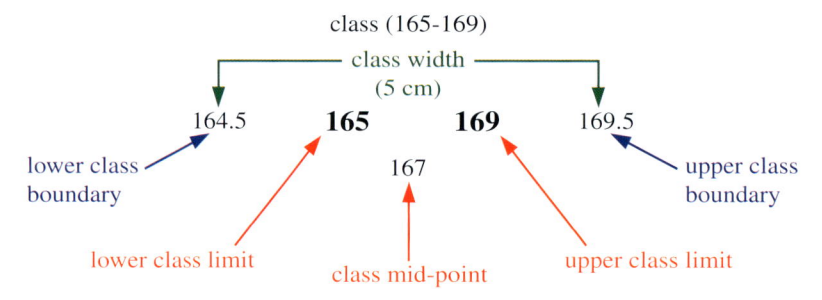

class (165-169)
class width (5 cm)
164.5 — lower class boundary
165 — lower class limit
167 — class mid-point
169 — upper class limit
169.5 — upper class boundary

Here is Tom's histogram. He has also drawn in a frequency polygon in green, joining the mid-points of the tops of the bars.

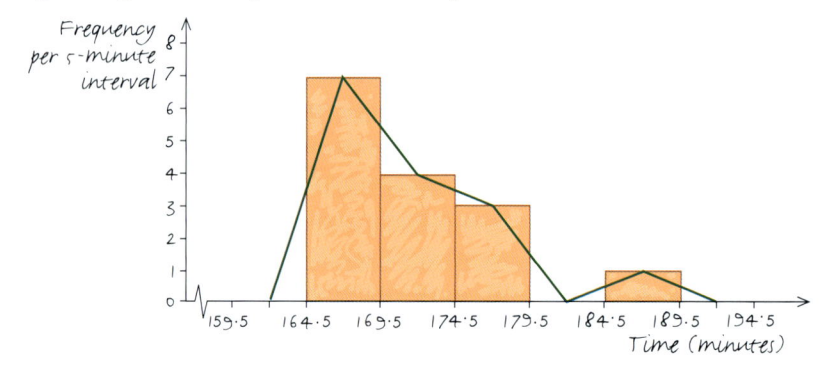

? Which display do you find more helpful, the histogram or the frequency polygon?

? The figures used in this example are the same as those on page 104, but rounded to the nearest whole number. Why are the histograms not the same?

10: Grouped data

1 The lengths (in minutes, to the nearest minute) of the phone calls made between Jill and Katharine are given in the table.

Time in minutes	1–5	6–10	11–15	16–20
Frequency	15	25	10	5

a) What are the boundary values for the class 6–10?

b) What is the mid-point of the class 6–10?

c) What is the longest possible time for a phone call in this table?

d) What is the shortest possible time for a phone call in this table?

e) Draw a histogram to show these data.

2 The weights (in grams, to the nearest gram) of the packs of information sent out one day by an investment adviser are given in the table.

Weight in grams	1–70	71–140	141–210	211–280
Frequency	350	420	560	79

a) What are the boundary values for the class 71–140?

b) What is the mid-point of the class 71–140?

c) What is the largest possible range for the data in this table?

d) Draw a histogram to show these data.

3 A group of 30 office workers went out for an office party. Their ages were as follows.

24	24	28	27	36	42	45	23	29	30
23	25	32	67	21	17	18	35	21	33
23	54	35	36	24	34	23	25	28	29

a) Construct a tally chart and frequency table for these data for the groups 16–20, 21–25, 26–30, 31–35, and so on.

b) Draw a histogram to show these data, taking care to label the ends of the intervals correctly.

c) The following Christmas the same people went out again. Describe how you could relabel your diagram.

4 On the opposite page, a frequency polygon has been added to the histogram of times.

a) Find the total area in the five rectangles of the histogram.

b) Find the area enclosed by the frequency polygon. What do you notice?

10: Grouped data

Frequency polygons

Chris Carter owns two clothes stores in Avonford. He wants one to appeal to young people and the other to older people.

To check this, he asks a sample of customers at each store to complete a questionnaire.

One of the questions asks for their age. Chris makes a frequency table of the ages at each store.

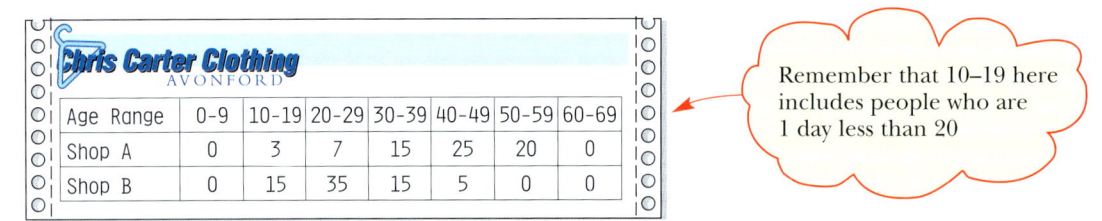

Chris Carter Clothing — AVONFORD

Age Range	0-9	10-19	20-29	30-39	40-49	50-59	60-69
Shop A	0	3	7	15	25	20	0
Shop B	0	15	35	15	5	0	0

Remember that 10–19 here includes people who are 1 day less than 20

One way to compare these data sets is to use a **dual bar chart**.

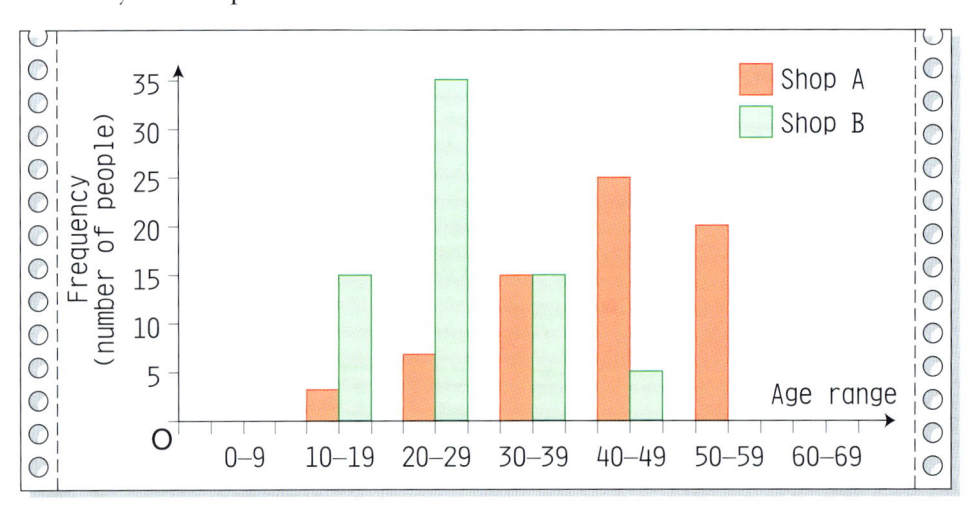

? *What does this chart tell you?*

Another way is to draw a **frequency polygon** for each set of data.

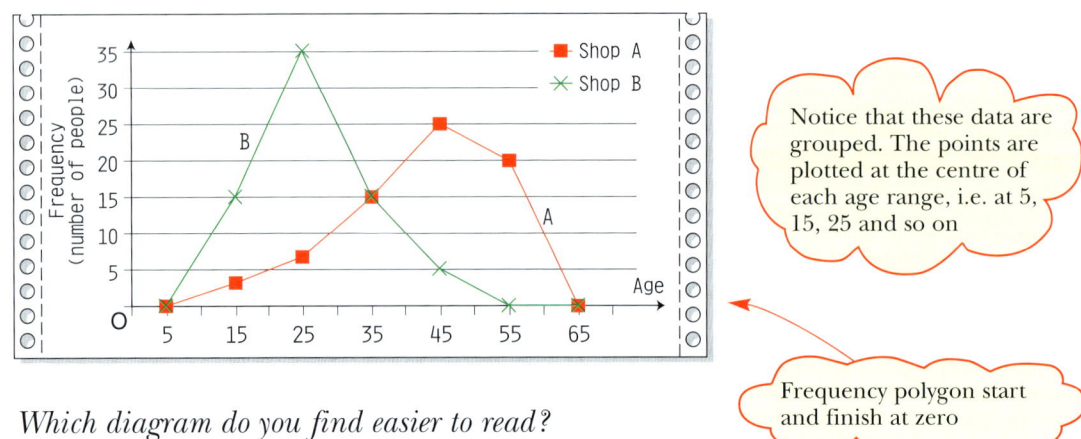

Notice that these data are grouped. The points are plotted at the centre of each age range, i.e. at 5, 15, 25 and so on

Frequency polygon start and finish at zero

? *Which diagram do you find easier to read?*

Is Chris Carter's policy working?

10: Grouped data

1 Two schools are competing in a cross-country race for teams of 40.

For each runner, the number of laps completed in one hour is recorded.

Results of cross-country race

Number of laps	3	4	5	6	7	8	9	10
Owens Academy	1	2	5	6	8	12	4	2
Avonford High	3	4	5	8	13	4	2	1

a) Draw a set of axes with number of laps (2 – 11) along the bottom and frequency (0 – 13) up the side.

b) Using these axes, draw a frequency polygon for each school.

c) Which school do you think performed better in the race?

2 Mark works for a coach company that runs day trips to various places of interest. The owner is keen to know the age profile and the gender of the customers so that she can decide where best to advertise each trip. Mark agrees to do a survey.

He presents the results from a trip to York in a dual bar chart.

a) What can you conclude from the chart?

b) Present the data in two frequency polygons on the same set of axes.

c) What can you conclude from your frequency polygons?

d) Which is easier to read, Mark's dual bar chart or your frequency polygon diagram?

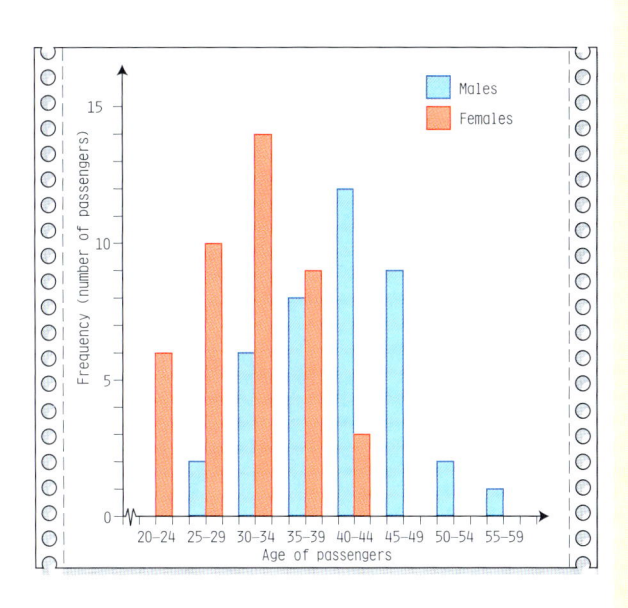

Measure the heights of equal numbers of boys and girls in your group.

Plot two frequency polygons on a large diagram to put on the wall.

What did you expect to find?

Were you right?

10: Grouped data

Mean, median and mode of grouped data

Debbie organises coach tours. She keeps a computer record of the age-group of each person booked on a particular tour. When she is ready to make the detailed arrangements for a tour she prints out a histogram like this one. She can then try to suit the activities and timings to their ages.

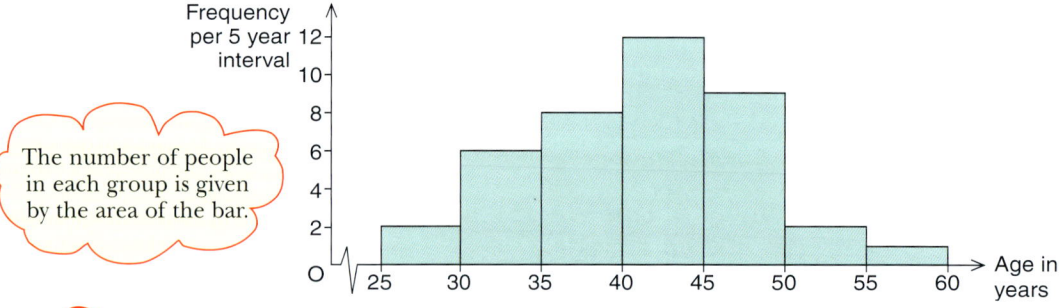

The number of people in each group is given by the area of the bar.

How many people on this trip are aged 25–29?

What is the modal age group?

How can you estimate the mean age of the people booked on this tour?

It is impossible to work out the mean of grouped data accurately. You need the raw data. However, you can make an estimate like this.

Age	20–24	25–29	30–34	35–39	40–44	45–49	50–54	55–59	
Mid-point of age group	22	27	32	37	42	47	52	57	
Frequency (number of people)	0	2	6	8	12	9	2	1	Total 40
Mid-point × frequency	0	54	192	296	504	423	104	57	Total 1630

This is the total number of people

The median of 40 ages is between the 20th and 21st ages. The median age group is 40–44

This is an estimate of the total age of this group - it assumes they are all 47

This is an estimate of the total age of all the people

Mean age of people = $\dfrac{\text{total age}}{\text{total number of people}}$

$\approx \dfrac{1630}{40}$

Mean age ≈ 40.75

Estimate the new mean when Frank Heys (aged 54) drops out of the tour.

10: Grouped data

1 Each passenger on a flight to Cairo is allowed one item of hand-luggage. The weights, to the nearest kilogram, of the hand-luggage items are given in the table.

Weight (kg)	1–5	6–10	11–15	16–20
Frequency (no. of items)	12	24	10	4

a) Which is the modal class?
b) What is the mid-point of the 6–10 kg class?
c) Estimate the total weight of hand-luggage on the plane.
d) Estimate the mean weight of the items.

2 Ben works for an estate agency which earns a percentage of the house price for each house it sells. He is working out the mean and mode of the house prices for the last year's sales, in order to prepare a cashflow forecast for the next year. He uses this table.

Price (£ thousands)	31-40	41-50	51-60	61-70	71-140
Frequency (no of houses)	35	42	56	7	3

a) Which is the modal class?
b) What is the mid-point of the class '71–140'?
c) Estimate the mean of these data.

3 In a recent cross-country skiing championship, the times for the first hundred competitors were recorded as follows.

Time (minutes)	80–84	85–89	90–94	95–99	100–104	105–109
Frequency (number of skiers)	8	27	33	20	8	4

a) Which is the modal class?
b) Estimate the mean time taken by these hundred skiers.

The fastest skier in this table was actually disqualified, so another skier whose time was 107 minutes entered the top hundred skiers.

c) Estimate the new mean time for the top hundred.
d) What is the new modal class for the top hundred?

Find the mean height of the people in your class or maths group.

Do it first using the individual heights, then from a grouped frequency table (if possible using a spreadsheet). Compare the answers you get by these two methods.

10: Grouped data

Cumulative frequency

There are 75 members of Avonford Squash Club. This table gives their ages grouped in 10 year intervals.

Age	0 – 9	10 – 19	20 – 29	30 – 39	40 – 49	50 – 59
Frequency (No. of members)	0	15	35	15	8	2

Another way to present these data is to use a **cumulative frequency table**. This gives the total number of people less than a given age.

Age (less than)	10	20	30	40	50	60
Cumulative frequency	0	15	50	65	73	75

These are the cumulative frequencies

50 members are aged under 30

0 + 15 + 35 + 15 = 65

You can plot the cumulative frequency on a graph like this:

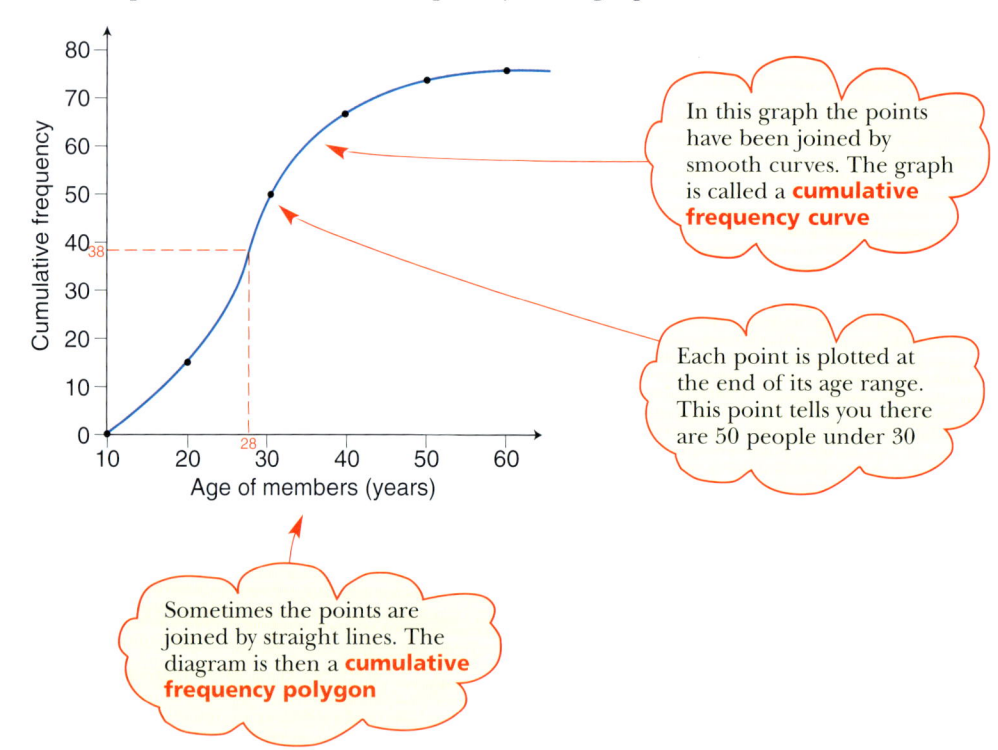

In this graph the points have been joined by smooth curves. The graph is called a **cumulative frequency curve**

Each point is plotted at the end of its age range. This point tells you there are 50 people under 30

Sometimes the points are joined by straight lines. The diagram is then a **cumulative frequency polygon**

You can estimate the median from this graph.

There are 75 people so the median is number 38.

Look at the red line on the graph. It goes along from 38 and down to the median value, which is 28 years.

Cumulative frequency curves are shaped like a sloping letter 'S'. Why?

112

10: Grouped data

1 This table shows the numbers of pupils of different ages attending school in Avonford.

In this table, 7–9 means at least 7 but not yet 9.

Age in years	Frequency (number of pupils)
4 – 5	300
5 – 7	550
7 – 9	650
9 – 11	670
11 – 13	640
13 – 15	590
15 – 17	580
17 – 19	220

a) Draw a cumulative frequency curve for these data.

b) Use your curve to estimate the median age of the school children.

c) Schools receive extra funding for pupils over 16. Estimate from your curve how many pupils in Avonford are over 16.

Using the data for the population of England on page 147, draw two cumulative frequency diagrams (on the same piece of graph paper), one for men and the other for women.

How do they show you that there are more elderly women than men?

2 The table shows the ages in completed years of the 70 partners in a ballroom dancing competition.

Age (completed years)	Number of men	Number of women
20 – 24	1	4
25 – 29	3	13
30 – 34	12	18
35 – 39	22	23
40 – 44	20	10
45 – 49	9	1
50 – 54	2	0
55 – 59	1	1

a) Using the same scales and axes, draw cumulative frequency curves for the ages of the men and the women.

b) From your curves, estimate the median age for the men and the median age for the women.

c) Compare the ages of the women and the men in the competition.

3 Many brown trout do not live for very long. The figures below show approximately how many out of 100 brown trout have died, and how many are still alive, at different ages. They are cumulative frequency figures.

Age	Number dead	Number alive
1 month	2	98
6 months	15	85
9 months	85	15
1 year	90	10
2 years	91	9
3 years	95	5

a) Plot these figures on the same graph.

b) What do you notice?

10: Grouped data

Quartiles

Dino has been keeping a record of the number of people who eat at his diner each evening. He wants to give a clear picture of his business to the bank manager.

He has drawn up a frequency table.

> This was Dino's sister's wedding party

No. of diners	5–9	10–14	15–19	20–24	25–29	30–34	35–39	40–44	45–49	50–54
Frequency	9	10	15	19	25	16	5	0	0	1

? *The greatest possible range of these data is 54 – 5 = 49.*

What is the smallest possible range?

Can you tell which of these is right?

Dino decides to make a cumulative frequency table.

> Remember that the cumulative frequency is the number of items *less than* a given value so it is plotted at the end of the interval

Number of diners (less than)	5	10	15	20	25	30	35	40	45	50	55
Cumulative frequency	0	9	19	34	53	78	94	99	99	99	100

He draws the cumulative frequency curve.

> Notice that the median has been taken as the 50th value, rather than half-way between the 50th and the 51st. It is usual to do this when the data set is large

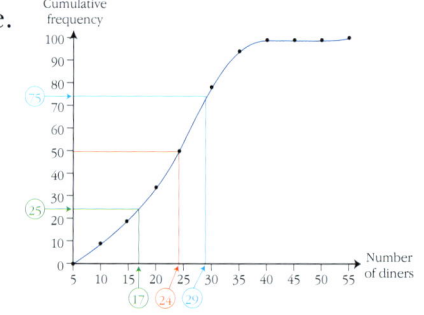

From his graph Dino can see that

- the median number of diners is about 24 (shown by the red line);

- on about 25 evenings (a quarter of the total) there were fewer than 17 diners. 17 is called the **lower quartile** and this is shown in green on the graph;

- on about 75 evenings (three quarters of the total) there were fewer than 29 diners. 29 is called the **upper quartile** and this is shown in blue on the graph;

- the difference between the upper and lower quartiles is 29 – 17 = 12. This is called the **inter-quartile range**.

? *On what proportion of the evenings did Dino have between 17 and 29 diners?*

? *Which is the better measure of spread in this case, the inter-quartile range or the range?*

10: Grouped data

1 The table shows the weight of breakfast cereal in 50 packets that are labelled as 375 g.

Weight of cereal (g)		Frequency (no. of packets)
At least	Less than	
355	360	2
360	365	4
365	370	17
370	375	20
375	380	4
380	385	2
385	390	1

These 2 packets are at least 355 g and less than 360 g

a) Plot these data on a cumulative frequency curve.

b) Use your curve to estimate the median and inter-quartile range of the weights. (Give your answers to the nearest whole number.)

c) Use your curve to estimate how many packets contain at least 376 g.

2 During an outbreak of chickenpox, Dr Rai had 200 patients who caught the disease. She kept a record of their ages as follows.

Age (in completed years)	0–9	10–19	20–29	30–39	40–49	50–59
Frequency (number of patients)	75	65	50	8	2	0

There are 65 people aged at least 10 but not yet 20

a) Draw a cumulative frequency curve for these data.

b) Estimate the median of the data from your graph.

c) Estimate the inter-quartile range from your graph.

d) Comment on your results.

Health visitors, school nurses and other health workers often carry growth charts. They check children's heights or weights against the charts to see whether they are within the expected range for their age.

Use a growth chart to find out the median height or weight for boys or girls of a particular age.

If possible, work out from the chart the lower quartile, the upper quartile and the inter-quartile range for that age.

10: Grouped data

Box-and-whisker diagrams (boxplots)

Diana is a biologist. The Forestry Commission have asked her to report on a plantation of pine tree saplings. She measures the heights of the trees and draws a cumulative frequency graph. To present a simple picture in her report, Diana also draws a box-and-whisker diagram.

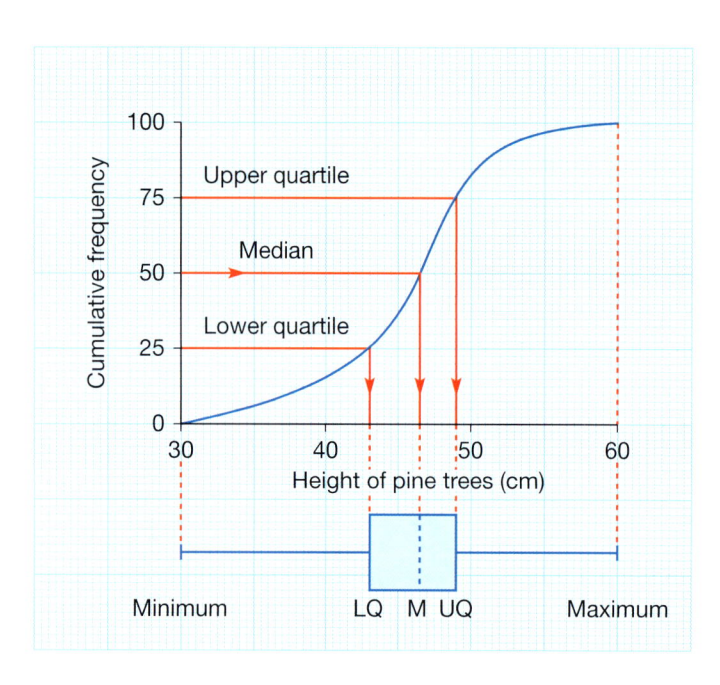

or boxplot

What are the main features of a box-and-whisker diagram?

What does the box show?

What do the whiskers show?

It is very easy to compare two (or more) distributions using boxplots. Look at these two. They show the marks in English and Maths of one year group in a school.

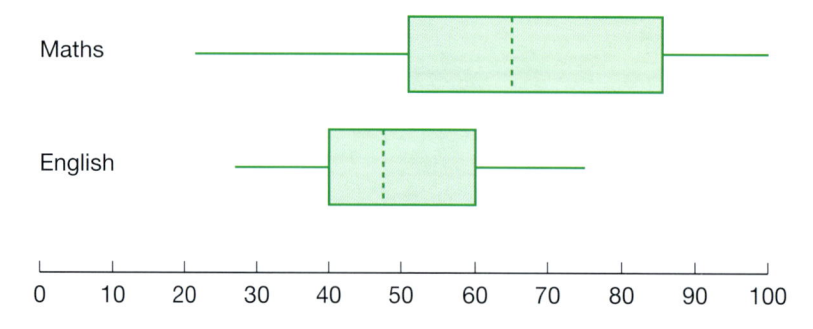

In which subject do most pupils get a higher mark?

In which subject do the marks cover a wider range?

116

10: Grouped data

Exercise

1 Find the median, mode and quartiles of the following values.

3 9 12 5 17 21 4 9 15 9 18

Draw a boxplot for these data.

2 A farm records the daily milk yield, in litres, of each cow in its herd. For one particular day the yields were as given below.

41 35 24 17 12 36 28 29 31 43 37
26 40 34 39 11 26 34 25 39 23 29
24 36 25 39 33 19 42 35 32

a) Draw an ordered stem-and-leaf diagram to show the data.

b) State the lower quartile, the median and the upper quartile.

c) Draw a boxplot for the data.

3 Jason carried out a survey of the ages of cars in a supermarket car park. Here is a frequency table of his results.

Age x (years)	$1 \leq x < 2$	$2 \leq x < 3$	$3 \leq x < 4$	$4 \leq x < 5$	$5 \leq x < 6$	$6 \leq x < 7$
Frequency	34	30	20	10	5	1

a) Compile a cumulative frequency table for these data.

b) Draw a cumulative frequency diagram.

c) Find the median and quartiles from your graph.

d) Draw the associated boxplot.

4 A company wanted to evaluate the training programme in its factory. They gave the same task to trained and untrained employees and timed each one in seconds.

Trained: 121 137 131 135 130 128 130 126 132
           127 129 120 118 125 134

Untrained: 135 142 126 147 145 156 152 153 149
           145 144 134 139 140 142

a) Draw a back-to-back stem-and-leaf diagram to show the two sets of data.

b) Find the medians and quartiles for both sets of data.

c) On the same scale draw the two boxplots.

d) Comment on the results.

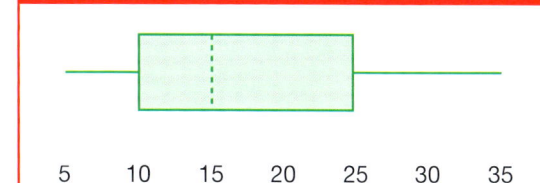

This boxplot shows the masses of 100 dogs in kilograms.

Draw the associated cumulative frequency graph.

10: Grouped data

Finishing off

Now that you have finished this chapter you should be able to

★ work out and use the mean, median, mode and range of a set of data
★ group data and draw frequency charts and histograms
★ draw and interpret frequency polygons
★ draw and interpret cumulative frequency curves and box-and-whisker diagrams
★ estimate the median and the inter-quartile range of a set of data from its cumulative frequency curve.

Use the questions in the next exercise to check that you understand everything.

Mixed exercise

1 I grew a row of 11 sunflowers in my garden. One day I recorded their heights in metres, as follows.

1.53   1.54   0.60   1.54   1.51   1.56   1.53   1.54   1.55   1.54   1.51

a) Write down the mode and the median.

b) Calculate the mean.

c) Give a reason for the difference between the mean and the median.

d) Which is the best average to use in this case?

2 A milkman delivers to 68 houses in one road. He keeps a record of the numbers of bottles of milk delivered.

a) Write down the mode.

b) Find the median.

c) Calculate the mean number of bottles delivered per house.

Juxon Street delivery

Number of bottles	1	2	3	4	5	6
Number of houses	24	21	8	13	0	2

3 This cumulative frequency curve shows the profits recorded by 100 companies.

a) Use the graph to produce a cumulative frequency table.

b) From the graph, estimate the median and quartiles.

c) Draw the associated boxplot.

d) Estimate the number of companies recording profits between £12 million and £22 million.

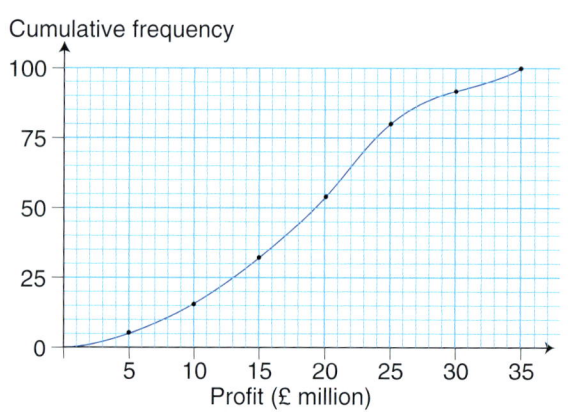

118

10: Grouped data

Mixed exercise

4 A machine is designed to pack tea into 200 g packets. To check on the machine's accuracy, a sample of packets is taken at regular intervals and weighed to the nearest gram. The results are shown in the table.

Weight of tea in grams	197	198	199	200	201	202	203
Frequency (no. of packets)	1	8	28	27	22	10	4

a) Find the total number of packets in the sample.

b) Explain why a weight of 199 grams means between 198.5 and 199.5 grams.

c) Estimate the mean weight of tea per packet.

d) State the modal weight.

e) Draw a frequency chart and a frequency polygon to illustrate these data.

f) State the range of weights.

g) Describe the machine's performance.

Here are the results from a sample taken one week later.

Weight of tea in grams	197	198	199	200	201	202	203	204
Frequency (no. of packets)	2	4	8	23	29	25	6	3

h) Draw the frequency polygon for these data on the same graph as for e).

i) Compare the two sets of data and describe what has happened to the machine.

5 The table gives the typing speeds of 160 entrants for a secretarial exam.

Speed (words/minute)	47 – 51	52 – 56	57 – 61	62 – 66	67 – 71	72 – 76	77 – 81
Frequency (no. of entrants)	16	50	32	24	22	10	6

a) Draw a cumulative frequency curve.

b) Estimate the median and the inter-quartile range.

c) Draw the associated boxplot.

d) Estimate the number of candidates gaining a distinction (with a speed greater than 75 words per minute).

e) A candidate needs to do 55 words per minute to pass. Estimate the number of candidates who failed the exam.

Eleven

Equations

Before you start this chapter you should

★ be familiar with the work in Chapter 3
★ be familiar with the work in Chapter 9.

Using equations

Mrs Singh uses this car hire company.
She pays them £140.

How many days' car hire does she get?

How did you get your answer?

HOLIDAY CAR HIRE
£20 fixed charge
+£15 per day

There are several ways to work out the answer to this question.
One of them is to use algebra to set up an **equation** and then **solve** it.

Start by letting n stand for the number of days. Then the cost is given by

$$20 + 15n = 140$$ *This is the equation*

Subtract 20 from both sides $\quad 20 + 15n - 20 = 140 - 20$

(Tidy up) $\quad\quad\quad\quad\quad\quad\quad\quad 15n = 120$

Divide both sides by 15 $\quad\quad\quad 15n \div 15 = 120 \div 15$

$$n = 8$$

Now we have solved the equation *Remember to do the same to both sides*

So Mrs Singh has the car for 8 days.

Check by substituting $n = 8$ in the left-hand side of the original equation:
$\quad 20 + 15 \times 8 = 140 \quad$ ✓ It is right.

Example Solve the equation $5x - 6 = 44$

Add 6 to both sides $\quad\quad\quad 5x = 44 + 6$ *Addition is the opposite of subtraction*

(Tidy up) $\quad\quad\quad\quad\quad\quad 5x = 50$

Divide both sides by 5 $\quad\quad x = 50 \div 5$ *Division is the opposite of multiplication*

$$x = 10$$

Some of the lines of this solution have been written more concisely than those in the example above. How?

Check by substituting $x = 10$:

$\quad 5 \times 10 - 6 = 44 \quad$ ✓ It matches the right-hand side, so it works.

The solution is $x = 10$.

11: Equations

1 Solve each of these equations, writing out the steps.

Check your answer by substituting it in the left-hand side of the original equation.

a) $n + 13 = 67$ b) $v - 4 = 11$
c) $3c - 5 = 19$ d) $2f + 15 = 99$
e) $3x + 7 = 37$ f) $2y - 4 = 16$

2 Solve each of these equations. Check your answers.

a) $2x + 4 = 10$
b) $3y - 5 = 13$
c) $7z + 100 = 135$
d) $5x + 12 = 37$

3 Solve each of these equations.

Check your answer by substituting it in the left-hand side of the original equation.

a) $4y - 16 = 24$ b) $5x + 12 = 24$ c) $5x + 7 = 17$
d) $3c - 4 = 8.3$ e) $2y - 8 = 12$ f) $10n + 4 = 13$
g) $x + 1 = 2$ h) $n - 42 = 36$ i) $12 + m = 23$
j) $13 + u = 16$ k) $2t + 5 = 25$ l) $5x - 6 = 4$
m) $3x - 2.6 = 4.9$ n) $12 + 2x = 14.6$ o) $7y + 4 = 4$

4 Make equations for the angles x, y and z shown in these diagrams.

Solve your equations to find the angles.

a) 120°, x

b) 40°, y, y

c) z, z, z, z, z

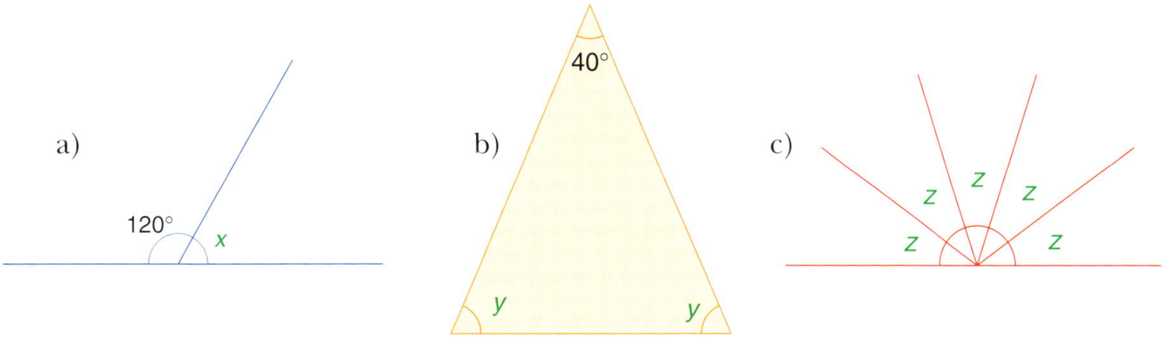

A company pays its employees a mileage rate when they use their own cars for work. This is 40p per mile for the first 100 miles and 30p per mile after 100 miles. Write this as two formulae for the payment £P for a journey of m miles. One fomula is for $m \leq 100$ and the other for $m > 100$.

Shomeet is paid £67 for one journey. Write down an equaion for m and solve it.

11: Equations

Solving equations

Can you think of a number which, if you multiply it by 6 then subtract 10, gives you the original number again?

You can write an equation to help you do this.

Call the number x. Then the equation is $6x - 10 = x$

> Multiplying x by 6 and subtracting 10 gives x

You can solve this using the same method as before.

Start with	$6x - 10 = x$
Add 10 to both sides	$6x = x + 10$
Subtract x from both sides	$6x - x = 10$
(Tidy up)	$5x = 10$
Divide both sides by 5	$x = 10 \div 5$
(Tidy up)	$x = 2$

> Add and subtract to get all the x terms on the left and the numbers on the right

> Multiply or divide once you have separated the x terms from the numbers

Check by substituting $x = 2$ in both sides of the original equation.

Left-hand side $= 6 \times 2 - 10 = 2$

Right-hand side $= 2$ ✓   Both sides are equal so $x = 2$ is correct.

> This is called **back-substitution**

The solution is $x = 2$.

Sometimes you need to solve equations with brackets in. To do this, you just expand the brackets and continue as before.

Example

Solve the equation $\quad 2(5 + x) = -2$

Solution

$2(5 + x) = -2$

(Expand the brackets)	$10 + 2x = -2$
Subtract 10 from both sides	$2x = -2 - 10$
(Tidy up)	$2x = -12$
Divide both sides by 2	$x = (-12) \div (+2)$
	$x = -6$

> In expanding the brackets we have just written the left-hand side differently: the right-hand side isn't affected

> Dividing a negative by a positive gives a negative

Check by back-substitution that $x = -6$ satisfies the original equation.

What is the solution of the equation $-5x = -40$?

11: Equations

1 Solve these equations. Write them out carefully and say what you have done at each step. Check your answers by back-substituting.

a) $h + 5 = 13$
b) $n + 3n = 12$
c) $5 + 2x = 22 + x$
d) $5y = 3 - y$
e) $k + 24 = 35 - 10k$
f) $2x - 7 = 19 - 2x$
g) $4x - 7 = 14 + 2x$
h) $8y + 1.2 = 7.2 + 3y$
i) $8x + 20 = 6x + 14$

2 Solve these equations. Check each answer by back-substituting.

a) $4 - 2x = 2$
b) $4 - 2a = 8$
c) $28 - 3x = 11x$
d) $22 - 6y = 13 + 3y$
e) $d + 20 = 4d$
f) $8 - 2x = 0$
g) $101 - 10k = 1 + 10k$
h) $1 - 10k = 101 + 10k$

3 Solve these equations. Check each answer by back-substituting.

a) $3(a - 2) = 2a$
b) $2(t - 3) = 8$
c) $4y + 2 = 3(y + 2)$
d) $11(c + 1) = 17 - c$
e) $5(x + 2) = 10 - x$
f) $17 - d = 9(1 - d)$

4 Equations do not always involve simple numbers, but so long as you know the method for solving them, all you need is a calculator to help. Try these, using a calculator when you need to.

a) $2.4c = 9$
b) $7y + 2 = 11$
c) $1.5 = 2.3 - 3x$
d) $9 + 99p = 333$
e) $4.5x = 13 - 2.5x$
f) $3.8y - 21 = 1.7y$
g) $6.2b + 3 = 5.5 + 4b$
h) $62 - 70a = 44 - 10a$

5 The formula for converting a temperature $F\,°$Fahrenheit into $C\,°$Celsius is given by

$$C = \frac{5}{9}(F - 32)$$

a) Find the value of C when $F = 68$
b) Find the value of F when $C = 30$
c) Copy and complete these steps to make F the subject of the formula.

Formula	$\frac{5}{9}(F - 32) = C$
$\times 9$	$5(F - 32) = 9C$
$\div 5$	$F - 32 =$
$+ 32$	$F =$

Investigation

One of these equations has no solution.

For the other equation, every value of x is a solution.

(i) $3(x + 6) - 2x = x + 18$

(ii) $5(2 + x) + 3x = 4(2x + 3)$

Which equation is which and how can you tell?

*When an equation is true for every value of x, it is called an **identity***

11: Equations

Equations with fractions

Sometimes you need to solve equations with fractions in them, such as

$$\frac{2}{3}x = 6$$

In this case, the number on the bottom of the fraction is 3 so you multiply both sides of the equation by 3:

$$3 \times \frac{2}{3}x = 3 \times 6$$

(Tidy up) $\quad 2x = 18$ ← *Now the fraction has gone*

Divide both sides by 2 $\quad x = 9$

? What is $\frac{2}{3}$ of 9?

Example Solve $\frac{3}{4}(x + 2) = x$

Solution

Multiply by 4 $\quad 4 \times \frac{3}{4}(x + 2) = 4x$

(Tidy up) $\quad 3(x + 2) = 4x$

(Expand brackets) $\quad 3x + 6 = 4x$

Subtract $4x$ and 6 $\quad 3x - 4x = -6$

(Tidy up) $\quad -x = -6$

Divide by -1 $\quad x = 6$ ← $(-6) \div (-1) = 6$

? *Substitute $x = 6$ in both sides of the equation. Does the check work?*

You use the same method to solve an equation when x is on the bottom line of a fraction. In this case you multiply both sides by x.

Example Solve $20 = \frac{360}{x}$

Solution

Multiply by x $\quad x \times 20 = x \times \frac{360}{x}$

(Tidy up) $\quad 20x = 360$

Divide by 20 $\quad x = 18$

This example could be about a pie chart with equal sectors of 20°.

The number of sectors is x. A more general formula is $A = \frac{360}{x}$ where $A°$ is the size of each sector.

? *Show how following the same steps gives the formula with x as subject,*

$$x = \frac{360}{A}$$

11: Equations

Exercise

1 Solve these equations.

a) $\dfrac{x}{2} = 5$ b) $\dfrac{x}{3} = 8$ c) $\dfrac{x}{100} = 100$

d) $\dfrac{x}{4} = -1$ e) $-\dfrac{x}{4} = -1$ f) $\dfrac{3}{4}x = 6$

Hint: multiply both sides by –1

2 a) Show that the equation $\dfrac{x}{12} + 2 = 5$ can be simpified to $\dfrac{x}{12} = 3$.

b) Solve the equation.

3 Solve these equations.

a) $\dfrac{x}{4} + 5 = 11$ b) $-\dfrac{2x}{3} + 8 = 6$ c) $\dfrac{5x}{4} - 5 = 10$

d) $\dfrac{1}{2}x + \dfrac{1}{2} = 6$ e) $\dfrac{1}{2}x + 6 = x$ f) $\dfrac{1}{3}x - 5 = \dfrac{2}{3}x - 8$

4 Solve these equations.

a) $\dfrac{1}{3}(x-5) = 2$ b) $\dfrac{1}{3}(x+5) = 1$ c) $\dfrac{1}{2}(x+7) = 4$

d) $\dfrac{1}{5}(x-1) = -2$ e) $\dfrac{2}{3}(x+2) = 0$ f) $\dfrac{2}{3}(x+\dfrac{1}{2}) = \dfrac{2}{3}$

g) $\dfrac{1}{2}(x+1) = x$ h) $\dfrac{3}{4}(x-2) = x-2$ i) $2(x-3) = \dfrac{1}{2}x$

5 Solve these equations.

a) $\dfrac{120}{x} = 8$ b) $\dfrac{15}{x} = 3$ c) $\dfrac{4}{x} = 8$ d) $\dfrac{9}{x} = -1$

6 The formula for the electric current I (ampères) in a circuit is $I = \dfrac{V}{R}$ where V is the voltage and R is the resistance in ohms.

a) In the case when $V = 240$ and $I = 6$

   (i) find R

   (ii) rearrange the formula to make R subject.

b) Write the formula with R as subject in the general case (i.e. in terms V and I).

Investigation

The diagram shows a sector of a pie chart with angle $x°$ where x is a whole number.

This can also be written as $y\%$ of the total. In some cases, y is also a whole number.

For example, $x = 180$ and $y = 50$.

What other whole-number values can x and y have?

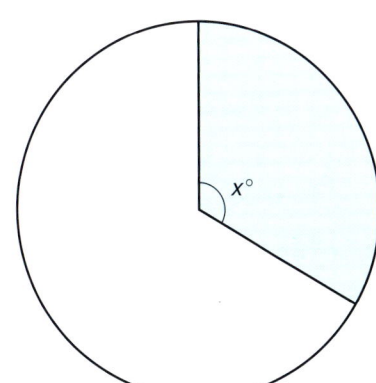

11: Equations

Changing the subject of a formula

Look at these two headlines. They both tell you the same thing.

Pele is the subject of the first. The winning goal is the subject of the second.

Look at this formula for finding the temperature, F, in degrees Fahrenheit from the temperature, C, in degrees Celsius.

$$F = 1.8C + 32$$

The Fahrenheit temperature F is on its own on the left. It is the **subject** of the formula. It is easy to find the value of F when you know the value of C.

Find the value of F when C is 20.

You probably found the formula very easy to use in this way. But what if you know the value of F and you want to find C?

Find the value of C when F is 77.

This would be easier with a formula for C in terms of F, in other words to have C as the subject of the formula. You can make C the subject by rearranging the formula as follows.

$$F = 1.8C + 32$$

Subtract 32 from both sides $\quad F - 32 = 1.8C$

Divide both sides by 1.8 $\quad \dfrac{F - 32}{1.8} = C$

or $\quad C = \dfrac{F - 32}{1.8}$

> Rearranging the formula is a bit like solving an equation. You want to get C by itself on one side, but at each step you have to do the same thing to each side of the formula

> This means the same as $(F - 32) \div 1.8$

> You might have seen this formula in another form:
> $C = \dfrac{5}{9}(F - 32)$
> They are both the same

The formula for the volume of a cylinder is $V = \pi r^2 h$. You can make r the subject by rearranging the formula.

$$V = \pi r^2 h$$

Divide both sides by πh $\quad \dfrac{V}{\pi h} = r^2$

Square root both sides $\quad \sqrt{\dfrac{V}{\pi h}} = r$

$$r = \sqrt{\dfrac{V}{\pi h}}$$

> Notice that $\sqrt{r^2} = r$
> In the same way,
> $(\sqrt{x})^2 = x$

11: Equations

1 Make *x* the subject of each of these.

a) $y = x + 4$
b) $y = x + 20$
c) $y = x + a$
d) $y = 3 + x$
e) $y = 13 + x$
f) $y = c + x$
g) $y = x - 5$
h) $y = x - 11$
i) $y = x - b$
j) $y = 6 - x$
k) $y = 1 - x$
l) $y = d - x$

2 Make *x* the subject in each of these.

a) $y = 2x$
b) $y = 0.1x$
c) $y = ax$
d) $y = \frac{x}{4}$
e) $y = \frac{x}{10}$
f) $y = \frac{x}{b}$
g) $y = \frac{3}{4}x$
h) $y = \frac{5}{3}x$
i) $y = \frac{a}{b}x$
j) $y = \frac{4x}{5}$
k) $y = \frac{11x}{2}$
l) $y = \frac{ax}{b}$
m) $p = x^2$
n) $q = 2x^2$
o) $l = x^2 - m$
p) $s = \sqrt{x}$

3 Make *t* the subject of each of these.

a) $x = 2t - 3$
b) $y = 3t + 4$
c) $p = 6 + 2t$
d) $c = 4 - t$
e) $z = 6 - 2t$
f) $s = 2t + a$
g) $x = 5t - c$
h) $n = 7t - 3x$
i) $p = t^3$

4 In each of these, make the given letter the subject.

a) $v = u + at$, u
b) $p = 2l + 2b$, l
c) $V = 4x - 9y$, x
d) $v = u + at$, t
e) $a = b + x^2$, x
f) $a = b + \sqrt{x}$, x

5 In each of these, expand the bracket and then make *x* the subject.

a) $p = 2(x + y)$
b) $V = 12(r + x)$
c) $s = 4(2 - x)$
d) $y = 4(a - x)$

6 In each of these, make the given letter the subject.

a) $A = lb$, l
b) $V = lbh$, h
c) $V = IR$, R
d) $c = \pi d$, d
e) $c = 2\pi r$, r
f) $I = \frac{r}{100} \times P$, P
g) $I = \frac{PRT}{100}$, T
h) $I = \frac{PRT}{100}$, R
i) $A = \frac{\pi d^2}{4}$, d

All of the formulae in question 6 are real.

What do they refer to?

Write down six more formulae that you can use in mathematics or elsewhere.

11: Equations

Finishing off

Now that you have finished this chapter you should be able to

★ solve simple equations using algebra

★ solve equations involving brackets and fractions

★ change the subject of a formula.

Use the questions in the next exercise to check that you understand everything.

Mixed exercise

1 Solve these equations.

a) $5a = 15$
b) $15b = 21$
c) $14 = 35c$
d) $2.5d + 11 = 17$
e) $67 = 16e - 13$
f) $22 - 3f = 7$
g) $14 - 5g = 1.5$
h) $22.4 - 4h = 10$

2 Solve these equations by collecting the terms in the unknown together on one side.

a) $5z - 4 = 2z + 5$
b) $6y + 1 = 7y - 3$
c) $8x + 9 = 15 - 4x$
d) $3w + 4 = 39 - 2w$
e) $12 + 6v = 2v + 30$
f) $14 + 5u = 26 - 3u$
g) $62 - 9t = 2t - 4$
h) $999 - 50s = 49s + 900$

3 Multiply out the brackets and solve these equations.

a) $2(8 + 2x) = 44$
b) $3(10 - x) = 5x$
c) $22x = 7(x + 3)$
d) $25 + x = 4(x - 2)$
e) $3(x + 2) = 2(x + 1)$
f) $5(2x + 4) - 3(x - 3) = 57$

4 Sunny Days Travel requires a deposit of 10% when a holiday is booked.

This can be written as a formula for the deposit £d for a holiday for n people priced at £p per person:

$$d = \frac{np}{10}$$

a) Calculate d when $n = 4$ and $p = 150$.

b) Make p the subject of the formula.

c) Jack paid a deposit of £85 for 2 people. What was the price per person?

5 All these equations have negative solutions. Solve them.

a) $3x = -15$
b) $-2x = 8$
c) $x + 7 = -13$
d) $2x + 9 = 3$
e) $2x - 5 = -11$
f) $4 - x = 6 + x$

11: Equations

Mixed exercise

6 Solve the following equations, which involve fractions.

a) $\dfrac{x}{3} = 6$

b) $\dfrac{x}{2} + 3 = 8$

c) $\dfrac{2x}{5} = 20$

d) $\dfrac{1}{4}(x + 3) = 2$

e) $\dfrac{1}{5}(x + 7) = 2$

f) $\dfrac{2}{3}(x + 8) = 6$

g) $\dfrac{42}{x} = 6$

h) $\dfrac{256}{x} = 32$

7 You can estimate the depth, d m, of a well by dropping a stone into it and finding t, the number of seconds before it hits the bottom. You then use the formula

$$d = 5t^2$$

Find the value of d when

a) $t = 1$

b) $t = 2$

c) Make t the subject of the formula.

d) Find t when d is 125 m.

d metres

8 The circumference, C, and area A, of a circle of radius r are given by the formulae

$$C = 2\pi r \quad \text{and} \quad A = \pi r^2$$

a) Find r in terms of C.

b) Find r in terms of A.

c) If $r = 12$, find C and A, leaving your answers in terms of π.

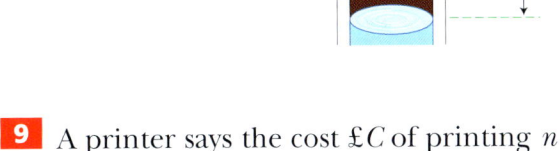

9 A printer says the cost £C of printing n greetings cards is given by

$$C = 25 + 0.05n$$

a) Find the cost of printing (i) 100 cards (ii) 1000 cards and (iii) 10 000 cards.

b) Find the cost per card in pence in each of the cases in a).

c) Find n in terms of C.

10 The Web electricity board has two ways of charging customers. The Silver price is £12 per quarter plus 6p per unit of electricity. The special Goldstar price is £36 per quarter plus 4.5p per unit.

Write down the total bill for x hundred units using

a) the Silver price

b) the Goldstar price.

c) Form an equation in x to find when the prices are equal. Solve it.

d) The Smiths use about 1400 units a quarter. Which price is cheaper for them?

The density d (grams per cubic centimetre) of a substance may be found by finding the mass m (grams) of a volume V (cubic centimetres) of it.

$$d = \dfrac{m}{V}$$

a) Write this formula with (i) m (ii) V as subject.

b) Conduct an experiment to find the density of various substances.

Twelve

Approximations

Decimal places

Connor is doing a survey to find out how well known leading twentieth century figures are. He asks 35 people to identify these two.

①  ②

Of the people he asks, 19 recognise ① to be Mother Teresa.

Connor works this out as a percentage.

$$\frac{19}{35} \times 100 = 54.2857\ldots$$

He has to decide how accurately to state this in his report.

How many decimal places, if any, do you think Connor should use?

Look at this number line.

54.2857...

|———|———|———|———|———|———|———|———|———|———|
54   54.5   55

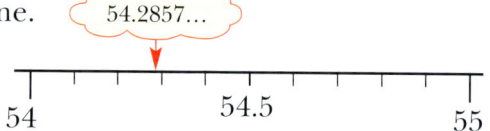

54.2857 . . . is nearer to 54 than to 55.
It is 54 correct to the nearest whole number.

Now look at this number line.

54.2857...

|———|———|———|———|———|———|———|———|———|———|———|———|———|———|———|———|———|———|———|———|
54.2   54.25   54.3   54.35   54.4

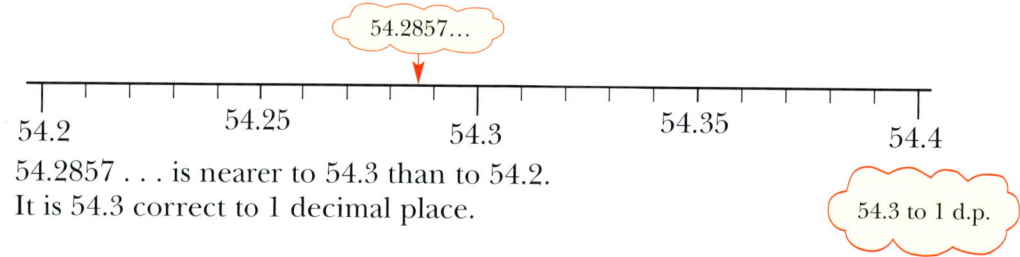

54.2857 . . . is nearer to 54.3 than to 54.2.
It is 54.3 correct to 1 decimal place.

54.3 to 1 d.p.

Copy the last number line. Colour the range of numbers which would be written as 54.3 correct to 1 decimal place.

Out of 35 people, 15 recognise ② to be Nelson Mandela.

What is this as a percentage

a) *correct to the nearest whole number?*
b) *correct to 1 decimal place?*

12: Approximations

1 The value of π is 3.141592 . . . Write this value correct to

    a) 1 decimal place   b) 2 decimal places   c) 3 decimal places.

2 Use your calculator to find the square root of 20.

    Write this value correct to

    a) 1 decimal place   b) 2 decimal places   c) 3 decimal places.

3 How many decimal places do you give if your answer is correct to the nearest

    a) tenth?   b) hundredth?

4 For each of these, write down your estimate of the reading to the nearest hundredth, and then write the reading correct to the nearest tenth.

    a)   b)

5 Write these fractions as decimals correct to 3 decimal places.

    a) $\frac{1}{6}$ b) $\frac{1}{12}$ c) $\frac{2}{3}$ d) $\frac{3}{7}$

6 Calculate 12.6% of 43.8, giving your answer correct to 2 decimal places.

7 a) Measure the length and width of this rectangle in centimetres giving your answers correct to 1 decimal place.

    b) Use these values to calculate the area of the rectangle giving your answer correct to 1 decimal place.

8 Work out the mean of these numbers giving your answer correct to 2 decimal places.

    4, 4, 5, 7, 8, 12, 15

9 A circle has radius 7.3 m. Using π = 3.14 calculate, correct to 1 decimal place,

    a) its circumference   b) its area.

Make the following measurements and decide how many decimal places it is sensible to use.

a) The height of a friend in metres.

b) The length and width of a sheet of paper in centimetres.

c) The length and width of a car in metres.

12: Approximations

Significant figures

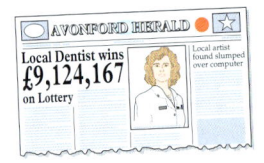

Look at these two newspaper headlines. Which do you think is better?

Sam has actually won £9 124 167, but the Avonford Star describes this as £9 million, which is £9 000 000. They decided their readers were not interested in the exact amount and so they rounded it to 1 **significant figure**.

£9 million	or £9 000 000	has 1 significant figure: it is the 9.
£9.1 million	or £9 100 000	has 2 significant figures: 9 and 1.
£9.12 million	or £9 120 000	has 3 significant figures: 9, 1 and 2.

Example

Write a) 294 217 to 3 significant figures (3 s.f.),

b) 0.004 297 to 2 significant figures.

Solution

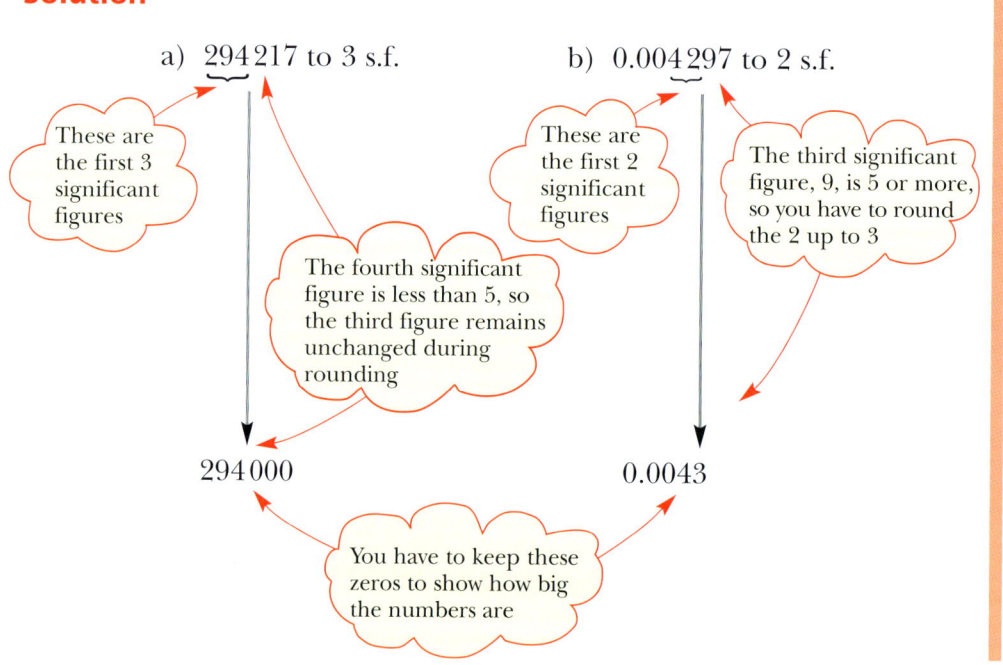

a) 294 217 to 3 s.f.

These are the first 3 significant figures

The fourth significant figure is less than 5, so the third figure remains unchanged during rounding

294 000

b) 0.004 297 to 2 s.f.

These are the first 2 significant figures

The third significant figure, 9, is 5 or more, so you have to round the 2 up to 3

0.0043

You have to keep these zeros to show how big the numbers are

In the example above, 0.004 297 is rounded to 2 significant figures.

What do you get if you round it to 2 decimal places?

Which way of rounding do you think is better?

Write 0.004 297 correct to 3 significant figures.

12: Approximations

1 Write each of these to the number of significant figures (s.f.) shown.

a) 77 328 to 2 s.f.
b) 42.195 to 3 s.f.
c) 5372 to 2 s.f.
d) 758 423 to 3 s.f.
e) 3780 to 1 s.f.
f) 61.977 to 3 s.f.
g) 6.7394 to 2 s.f.
h) 53 660 to 1 s.f.
i) 33.830 7 to 3 s.f.
j) 0.005 38 to 1 s.f.

2 For each of these, write down your estimate of the reading to 4 significant figures, then write the reading correct to 3 significant figures.

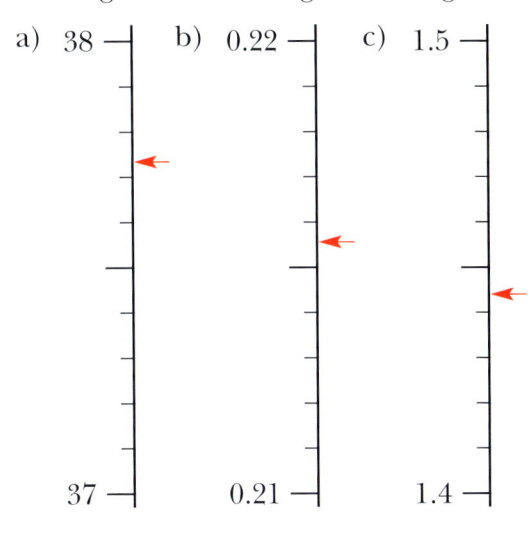

a) 38 — 37
b) 0.22 — 0.21
c) 1.5 — 1.4

3 Repeat question 2 giving your answers correct to 2 significant figures.

4 Work out 15% of 175 000 correct to 2 significant figures.

5 A football pitch is 114 m long and 73 m wide. Work out, correct to 2 significant figures

a) the perimeter
b) the area.

6 The profits of a business for the four quarters of a year are

£55 000   £79 000
£127 000   £98 000

a) Work out the average quarterly profit correct to 2 significant figures.
b) Work out the total profit for the year correct to 2 significant figures.

7 Tickets for a play are £4.25 each. This table shows the number of tickets sold for each of the three performances.

Day	Thursday	Friday	Saturday
Tickets sold	257	319	348
Income			

a) Copy and complete the table, writing each day's ticket income correct to 3 significant figures.
b) Write down the total income correct to 2 significant figures.

Measure the length and width of a page of this book in centimetres. Use your answers to work out the area of the page.

What is a sensible number of significant figures for your answer?

12: Approximations

Estimating costs

Ross is buying these items. He wants to estimate the total cost before going to the checkout.

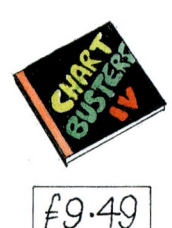

£5·45   £9·49   £10·99

? *Without using a calculator and without writing anything down, estimate the total cost.*

You may have worked out the total cost by rounding each amount to the nearest pound, then adding them:

£5 + £9 + £11 = £25

You may have rounded each amount to the nearest 50p and then added them:

£5.45 is about £5.50   £9.49 is about £9.50   £10.99 is about £11.00

£5.50 + £9.50 + £11.00 = £26.00

? *Which of these is the better estimate?*

Since Christmas, Ross has bought 4 posters, 4 compact discs and 2 T-shirts.

Ross, you must have spent nearly £100 at that shop

No mum, honest — it can't be more than £60

They sit down together and estimate how much Ross has spent.

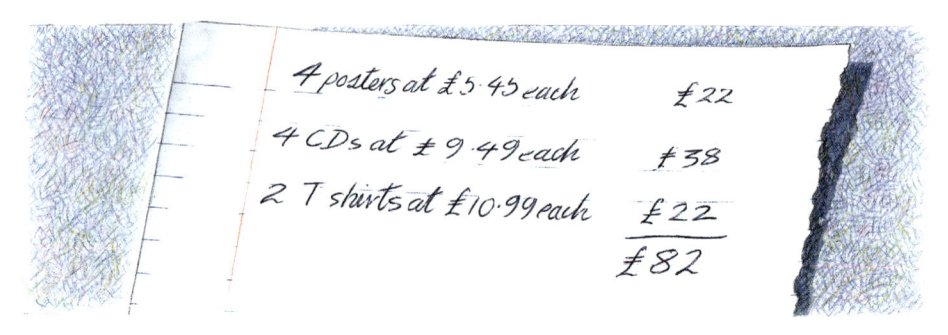

```
4 posters at £5.45 each        £22
4 CDs at £9.49 each            £38
2 T shirts at £10.99 each      £22
                               ----
                               £82
```

? *What is the exact cost?*

12: Approximations

1 Estimate the cost of each person's bill.

a) Anna buys 2 sunloungers.

b) Mitchell buys a table and a parasol.

c) Jill buys a sunlounger, a parasol and a lilo.

d) Wilf buys 4 sunhats and a lilo.

e) Jake buys a table and 4 chairs.

f) Eleanor buys a table, 3 chairs, a parasol and a sunlounger.

Sunhat £4.99, Chair £34.50, Table £24.75, Parasol £10.99, Sunlounger £19.45, Lilo £4.99

2 Five 42-seater coaches are almost full for a sight-seeing tour.

a) Estimate the total number of tourists in them.

b) Each tourist has paid £9.50 for the tour.

Estimate the total ticket income.

3 'Mega' carpet tiles are 1 metre square and cost £3.95 each. Estimate the cost of covering a rectangular area 4.9 m by 3.85 m.

4 Estimate each person's bill at this hotel.

Bed & Breakfast	£21 pppn
Evening Meal	£8·50 extra

a) William books bed and breakfast for 4 nights.

b) Suja books bed, breakfast and evening meal for 2 nights.

c) Linda and Pete book bed and breakfast for 6 nights.

d) Ian books bed, breakfast and evening meal for 5 nights for 4 people.

Fish £3.75, Milk £1.49, Chicken £5.65, Biscuits £4.25, Cheese £1.29, Coffee £2.59

Hana buys these six items. She estimates the cost by first combining items that add up to roughly a whole number of pounds.

She notices that the fish and the biscuits together cost £8. The milk and the coffee come to about £4. The chicken and the cheese come to about £7.

Hana estimates the total bill as

£8 + £4 + £7 = £19

Find a supermarket bill with between 10 and 15 items on it and see if you can make a rough estimate of the cost in a similar way.

Stick the bill on a sheet of paper and show your working alongside.

12: Approximations

Using your calculator

You should always check that answers you get on your calculator are sensible.

Mr Harris wants to reseed his lawn. He asks his 4 children to work out the area of the lawn.

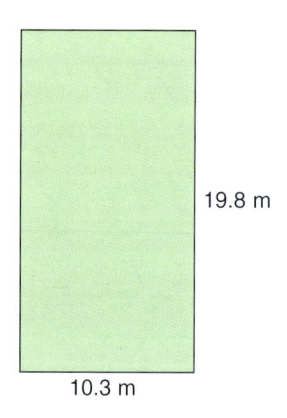

19.8 m

10.3 m

They each try to work out 19.8 × 10.3 on their calculators. Here are their answers.

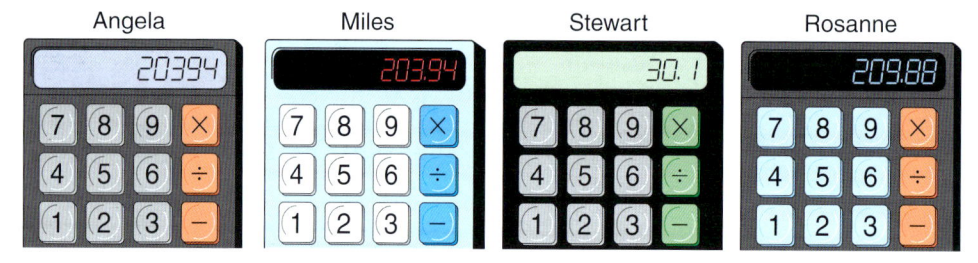

Angela: 20394
Miles: 203.94
Stewart: 30.1
Rosanne: 209.88

Which of these answers seem sensible?

One way to decide is to say 19.8 is about 20 and 10.3 is about 10, so the answer should be roughly 200.

Miles and Rosanne both have sensible answers. It is not easy to judge which of them is correct without doing an accurate calculation, but Mr Harris might decide that 200 is accurate enough for his purpose.

Do the calculation on your calculator and find out which is correct.

What mistake did Angela make?

What mistake did Stewart make?

Can you tell what mistake Rosanne made?

Rosanne actually keyed in 10.6 instead of 10.3 by mistake.

As it only made a small difference to her answer, a rough check did not show up the fault. However, by doing his rough check Mr Harris did avoid getting the answer completely wrong.

Lawn seed costs £8.99 for a packet that will cover about 100 m^2.

What is the cost of reseeding Mr Harris's lawn?

12: Approximations

Do the questions in this exercise quickly, using your calculator. Then check that none of your answers is silly by doing rough calculations by hand.

1 Estimate the area of these shapes.

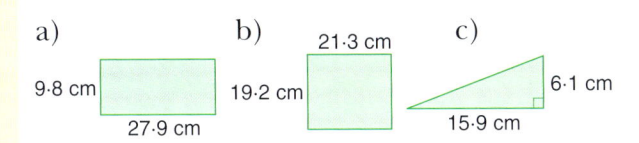

a) 9·8 cm by 27·9 cm

b) 21·3 cm by 19·2 cm

c) triangle, 15·9 cm and 6·1 cm

2 Geoff employs 11 people who earn between £180 and £220 each per week. Estimate

a) Geoff's total wage bill for a week;

b) Geoff's total wage bill for a year.

3 Marcus is touring the world. He changes £80 into the local currency in each country he visits.

Estimate how much he gets in

a) Malaysia
b) Thailand
c) Fiji
d) Australia

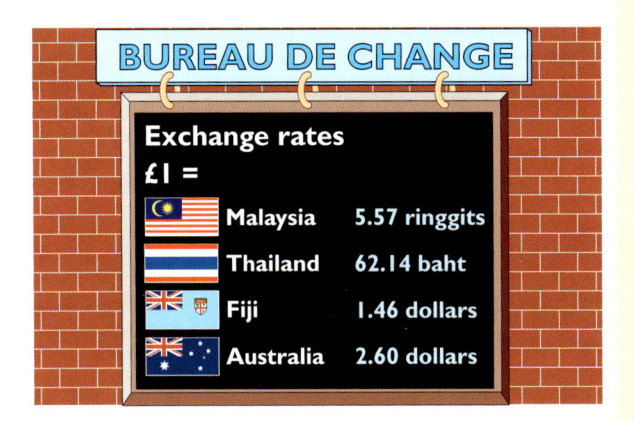

BUREAU DE CHANGE

Exchange rates
£1 =
Malaysia   5.57 ringgits
Thailand   62.14 baht
Fiji   1.46 dollars
Australia   2.60 dollars

4 One Saturday the attendances at five football grounds in one county are

7927   26 371   9925   12 304   13 849

a) Estimate the average attendance.

b) In a season each team plays 21 home games. Estimate the total season's attendance at each ground.

c) Suggest a reason why the figures in b) may be poor predictions.

5 Estimate the perimeter of each of these shapes. (For rough calculations like this you can take π as 3.)

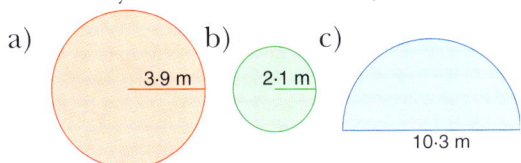

a) circle, 3·9 m

b) circle, 2·1 m

c) semicircle, 10·3 m

6 Estimate the area of each shape in question 5. (Take π as 3 again.)

Estimate the dimensions of a typical public swimming pool and draw suitable diagrams to illustrate its shape. Estimate the volume of water it contains.

137

12: Approximations

Using fractions and percentages

Elliott estimates the saving on these trainers.

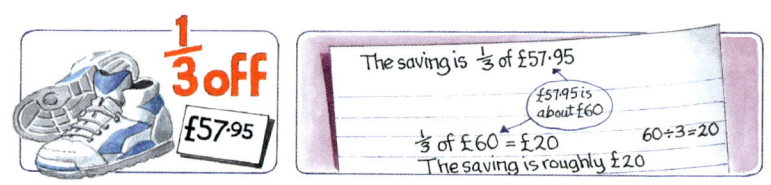

1/3 off £57·95

The saving is 1/3 of £57·95
£57·95 is about £60
1/3 of £60 = £20
60 ÷ 3 = 20
The saving is roughly £20

? Estimate the saving on trainers priced at £89.45 and marked '$\frac{1}{3}$ off'.

Helena has done a survey of 600 people to find out about their housing.

One of the questions was about when the property was built.

She presents the results in this pie chart.

Without measuring any angles, you can work out roughly how many of the people live in each age of property.

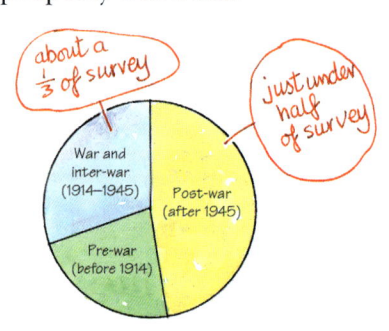

about a 1/3 of survey — War and inter-war (1914–1945)
just under half of survey — Post-war (after 1945)
Pre-war (before 1914)

Just under half of them live in post-war properties:

$\frac{1}{2}$ of 600 = 300

so just under 300 live in post-war properties.

About one third live in war and inter-war properties:

$\frac{1}{3}$ of 600 = 200

so about 200 live in war and inter-war properties.

Remember that $\frac{1}{2}$ of 600 is the same as $\frac{1}{2} \times 600$

? Roughly how many live in pre-war properties?

Bridget earns £13 900. She receives a pay rise of 9.2%. She makes this rough estimate of how much extra she will earn.

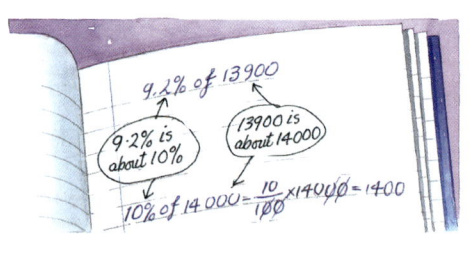

9.2% of 13 900
9.2% is about 10%
13 900 is about 14 000
10% of 14 000 = $\frac{10}{100} \times 14000 = 1400$

? Use a calculator to work out Bridget's new salary. How do the results compare?

Bridget rounded both numbers (the 9.2 and the 13 900) upwards.

How could she have found a better estimate?

12: Approximations

1 Estimate the answers to these calculations.

a) $\frac{1}{2}$ of 407   b) $\frac{1}{4}$ of 99   c) $\frac{1}{10}$ of 1793   d) $\frac{1}{3}$ of 595

e) $\frac{1}{6}$ of 1492   f) $\frac{3}{4}$ of 810   g) $\frac{2}{3}$ of 460   h) $\frac{3}{10}$ of 1187

2 A travel company sold 30 273 holidays last year.

They know that 1 in 3 holidaymakers went to Spain, 1 in 4 to France and 1 in 6 to Greece.

Estimate the number of holidays the company sold to

a) Spain   b) France   c) Greece.

3 This pie chart shows the distribution of a sports grant of £11 945.

Estimate the amount given to

a) athletics   b) swimming
c) badminton   d) golf.

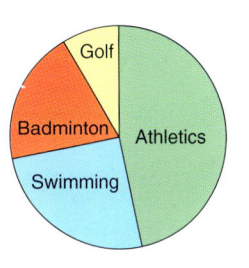

4 Estimate the answers to these calculations.

a) 51% of 607   b) 23% of 101

c) 48% of 395   d) 33% of 121

e) 49% of 362   f) 27% of 966

g) 73.5% of 197   h) 9.7% of 587

5 Estimate the salary increases for

a) Philippa who earns £20 376 and gets a 3.9% rise;

b) Simon who earns £9980 and gets a 2.7% rise;

c) Louise who earns £14 360 and gets a 4.85% rise.

6 Estimate the sale price of

a) the shirt   b) the jumper   c) the trainers.

(shirt £15.95, jumper £34.75, trainers £58.95, 25% OFF ALL MARKED PRICES)

Find some examples of survey results in newspapers or magazines. Are they given as simple fractions, percentages, in pictorial form or in some other way?

Write a short report about what you find.

12: Approximations

Errors

Some errors always occur when numbers are rounded. Julian describes the top of his coffee table as 44 cm by 32 cm. Both of these measurements are given to the nearest centimetre

44 cm means between 43.5 cm and 44.5 cm
32 cm means between 31.5 cm and 32.5 cm

So the area of the top of Julian's coffee table lies between 31.5×43.5 and 32.5×44.5 cm^2. That is between 1370.25 and 1446.25 cm^2.

What can you say about the perimeter of Julian's coffee table?

Julian measures the table very carefully and finds it is 44.1 cm by 32.1 cm, or 441 mm by 321 mm.

Using these measurements the area is $44.1 \times 32.1 = 1415.61$ cm^2, or 141 561 mm^2.

What is the perimeter?

These measurements are to the nearest millimetre.

What are the greatest and least values for the length?

What are the greatest and least values for the width?

What are the greatest and least values for the area?

What are the greatest and least values for the perimeter?

There is an actual error of 0.1 cm or 1 mm between Julian's description and his measurements. The error was the same for the length and the width, but it is a larger proportion of the width.

12: Approximations

Exercise

1 Ben Faolin has a height of 3900 ft to the nearest 100 ft.

Find the greatest and least values of the height of Ben Faolin.

2 Ian's car does 45 miles to the gallon measured to the nearest 5 miles. His petrol tank holds 7 gallons.

Find the greatest and least distances that Ian can expect to travel on a full tank of petrol.

3 Julie is moving to a new house. She measures the staircase and finds that there are 15 steps, i.e. 15 riser and 14 treads.

Each riser is 18 cm high to the nearet cm and each tread is 23 cm long to the nearest cm.

 a) Find the greatest and least heights of each riser.

 b) Find the greatest and least lengths of each tread.

 c) Find the greatest and least amounts of carpet that Julie will need to cover the staircase.

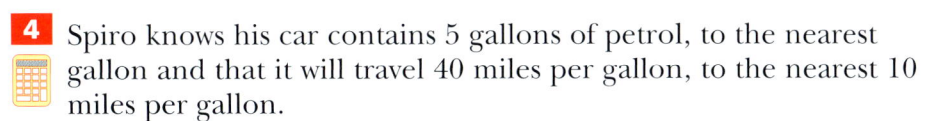

4 Spiro knows his car contains 5 gallons of petrol, to the nearest gallon and that it will travel 40 miles per gallon, to the nearest 10 miles per gallon.

 a) What is the greatest distance Spiro can hope to travel without running out of petrol?

 b) What is the least distance Spiro can be certain to travel without running out of petrol?

5 A bird keeper describes her aviary as 15 metres by 10 metres by 3 metres, with each measurement to the nearest 1 metre.

State the smallest and largest possible volumes of the aviary.

Here are some well known approximate measures

> Tip of your thumb to its first joint : 1 inch
> Span of your hand when spread out : 6 inches
> 1 pace: 1 metre (or 1 yard)
> Tips of fingers to chin (when your arm is outstretched) : 1 metre (or 1 yard)

Measure each of these on yourself and estimate the percentage errors using the measures.

12: Approximations

Finishing off

Now that you have finished this chapter you should be able to

★ round to a given number of decimal places

★ round to a given number of significant figures

★ estimate costs

★ check that the answers on your calculator are sensible

★ make rough calculations

★ understand the greatest and least values of a rounded number.

Use the questions in the next exercise to check that you understand everything.

Mixed exercise

1 Use your calculator to find the square root of 28.

Write this value correct to

a) 1 decimal place   b) 2 decimal places   c) 3 decimal places.

2 Estimate each of these readings to 2 decimal places.

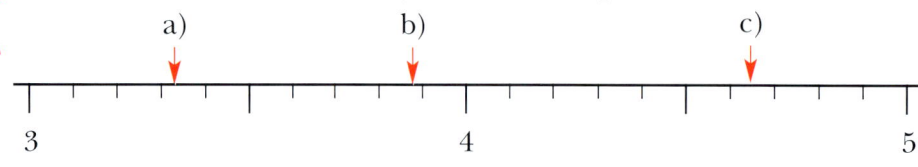

3 Work out the area of each shape giving your answer correct to 1 decimal place.

a) 2·74 m by 5·82 m rectangle

b) triangle with 6·84 m and 9·53 m

c) circle with 5·7 m

4 Write each of these to the number of significant figures (s.f.) shown.

a) 8397 to 1 s.f.

b) 764 729 to 3 s.f.

c) 14.7528 to 3 s.f.

d) 0.06527 to 2 s.f.

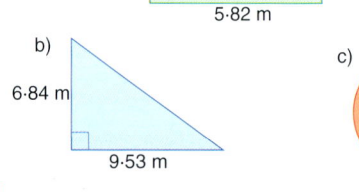

5 This bar chart shows the amount of money raised by each of the four houses at Westfield School.

a) Estimate, correct to 2 significant figures, the amount raised by each house.

b) Estimate the total amount raised correct to 2 significant figures.

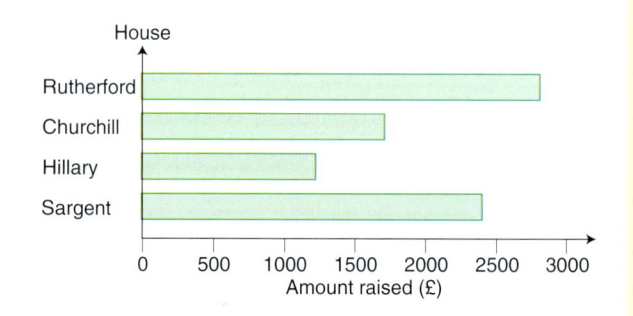

142

12: Approximations

Mixed exercise

6 Estimate the cost of each person's bill from this takeaway.

a) Monika orders cod and chips twice and 2 colas.

b) Chris orders chicken and chips twice and 2 small pizzas.

c) Emma orders meat pie and chips, 2 large pizzas and 3 colas.

MENU			
Cod	£1.99	Pizza (small)	£3.95
Chicken	£2.85	Pizza (large)	£4.95
Meat Pie	£1.10	Cola	£0.99
Chips	£0.99		

7 Kitchen tiles are 20 cm by 20 cm and cost £1.89 each. Lauren tiles an area 195 cm by 55 cm. Estimate the total cost of the tiles.

8 Estimate the answers to these calculations.

a) 9.4×30.2
b) $8.092 + 3.95$
c) $407.8 \div 4.1$
d) 19.7^2
e) $\sqrt{63.41}$
f) $26\,172 - 8\,395$
g) $\frac{1}{4}$ of 843
h) $\frac{2}{3}$ of 913
i) 74% of 982

9 Rio wants to returf his lawn. It is 8.1 m long and 5.8 m wide. The turf costs £7.95 per square metre.

Estimate the cost of returfing the lawn.

10 The length of a rectangular room is 8 m, to the nearest metre. Its width is 6 m, to the nearest metre.

a) Find the smallest and largest possible values for its perimeter.

b) Find the smallest and the largest possible values for its area.

11 VAT at 17.5% increases the total cost of an item by about a sixth.

Use this fact to estimate the cost (including VAT) of these.

a) £720 ex VAT

b) £149 ex VAT

c) £475 ex VAT

Estimate the cost of a wedding reception, a week's youth hostelling, or buying food for your family for one week.

Thirteen

Practical statistics

Before you start this chapter you should

★ be familiar with the work in Chapter 7

★ be familiar with the work in Chapter 10

★ be able to draw and interpret a pictogram.

Use the questions in the next exercise to check that you still remember these topics.

Revision exercise

1 The following data are collected for a group of students. State whether they are categorical or numerical. If they are numerical say also whether they are continuous or discrete.

a) their heights, measured in cm
b) their weights, measured in kg
c) the days of the week on which they were born
d) their star signs
e) the number of brothers and sisters they have.

2 This pictogram illustrates the lengths of the races at a meeting, and the number of horses in them.

a) State the length of each race, and the number of horses in it.
b) Which race has the most horses?
c) Which is the shortest race?

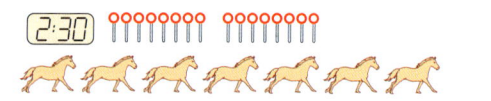

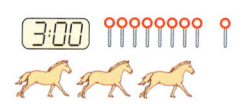

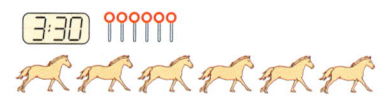

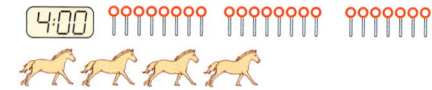

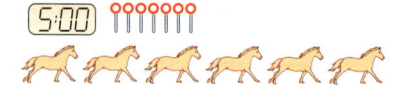

13: Practical statistics

Revision exercise

3 Within England there are 45 000 hectares of derelict land. It has become derelict for a range of reasons. The percentage of derelict land attributed to different causes is shown in this table.

Cause	Slag heaps	Excavations	Disused railway lines	Military	Other
Percentage	30	20	23	10	17

a) Work out the area of land attributed to each cause.

b) Draw a bar chart showing the areas you found in a).

c) It is estimated that a farmer can rear 12 sheep per hectare of derelict land. How many sheep could be reared on the disused railway lines?

4 During one day 30 people go sky diving at a club. Their ages are as follows.

```
23  25  32  67  21  17  18  35  21  33
23  54  35  36  24  34  23  25  28  29
24  24  28  27  36  42  45  23  29  30
```

a) Construct a tally chart and a frequency table for these data for age groups 16–20, 21–25, 26–30, 31–35 etc.
b) Which age group has the most sky divers?
c) A newspaper reporter is present and writes the headline 'Three generations sky dive together'.
Look at the figures and decide if this is possible.

5 A small country sends a squad of 15 athletes to the Olympic Games. Each athlete only enters one event. 3 win gold medals, 1 gets a silver and 4 get bronze. The rest get no medals.

a) Draw a bar chart to illustrate the results.
b) What percentage of athletes win a medal?

You are going to collect customer data for a travel agent, supermarket, health club, or other business of your choice.

Decide on several easily-recognisable categories, e.g. male, female, different age groups, alone, couples, or families.

Choose an appropriate place and collect data.

Collate your data into frequency tables and keep it carefully to use later. Several of you could pool your information and set up a database to do the collation.

13: Practical statistics

Reminder

In a set of numbers

- the **mode** is the number that occurs most frequently
- the **mean** is commonly called the average: it is found by dividing the total of the numbers by how many there are
- the **median** is the value of the middle number when they are placed in order. (For an even number of numbers, the median lies half-way between the middle two)
- the **range** is the difference between the highest number and the lowest number.

Use the questions in the next exercise to check that you still remember these topics.

Revision exercise

1 Find the mean, median, mode and range for each of these data sets.

a) 14, 10, 3, 9, 8, 8, 11

b) 4, 4, 5, 6, 8

c) 1, 1, 3, 4, 2, 6, 2, 4, 3, 2

d) 10, 10, 10, 10, 11, 33

2 a) The mode of the numbers 1, 1, 3, 9, 9, x, 7 is 1.
What is the value of x?

b) The median of the numbers 4, 6, 8, x, 11, 14 is 9.
What is the value of x?

c) The mean of the numbers 3, 5, 7, 10, 14 and x is 8.
What is the value of x?

3 Greg works out that his mean monthly salary for the past year has been £606. He got £600 for each of the first 11 months.

What was his salary for December?

13: Practical statistics

Revision exercise

4 Here are the heights, in metres, of a group of people.

1.6, 1.7, 1.9, 1.8, 1.6, 1.7, 1.5, 1.9, 1.6, 1.8

a) What is the modal height?

b) What is the median height?

c) What is the mean height?

d) When a new member joins the group, the mean height becomes 1.7 m exactly. How tall is the new member?

e) How does the newcomer affect the mode and the median of the data?

5 Jacob has done a survey to find out how many people live in the houses in his street.

He has produced this data collection sheet.

a) Make a frequency table to show his data.

b) What is the range of the data?

c) What is the mode of the data?

d) Draw a vertical line chart of the data.

e) How can you find the mode from the vertical line chart?

No. of occupants	No. of houses
0	/
1	////
2	//// //
3	////
4	///

The table shows the expected age distribution of the population in England in 2011, in millions. The total population is expected to be 52 000 000 people.

	0–14	15–29	30–44	45–59	60–74	75 and over	Total
Males	4.7	5.1	5.4	5.4	3.7	1.6	
Females	4.3	4.9	5.2	5.3	4	2.4	
People			10	10.6	10.7	7.7	

Source: Office for National Statistics

Copy and complete the table.

Are there more males or females in England?

Find the percentage of under-15s who are a) male b) female.

Find the percentage difference between the numbers of males and females who are under 15.

Similarly find the percentage difference between the numbers of males and females who are 75 or over. Comment.

13: Practical statistics

Which average?

Data are often given in a frequency table, as in the example below.

Fifty young adults are asked how many days they worked during the last week.

Number of days	0	1	2	3	4	5	6	7
Frequency (number of people)	10	0	2	5	9	19	5	0

- The mode is 5 days. This is the number with the greatest frequency.
- The median of 50 numbers is midway between the 25th and 26th numbers.

    You don't actually need to write them all out. You can think of them like this:

    0, ..., 0, 2, 2, 3, ..., 3, 4, ..., 4, 4, 5, ..., 5, 6, ..., 6
    1 ... 10 11 12 13 ... 17 18 ... 25 26 27 ... 45 46 ... 50
                                        ↑
    The median is here; it is 4.

- The mean is $\dfrac{\text{total number of days worked}}{\text{total number of people}}$

    $= \dfrac{0 \times 10 + 1 \times 0 + 2 \times 2 + 3 \times 5 + 4 \times 9 + 5 \times 19 + 6 \times 5 + 7 \times 0}{10 + 0 + 2 + 5 + 9 + 19 + 5 + 0}$

    $= \dfrac{180}{50} = 3.6$

Which of these is the best average? That depends on why you want to know it.

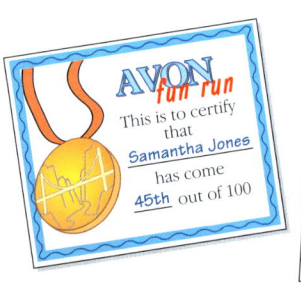

"I am better than average." Her average is the median.

In this case the average is the mean.

In this case the average is the mode.

These are the ages of people on a school trip to France.

Age	14	15	45	51
Frequency	8	5	2	1

Which is the best average to use to describe them?

13: Practical statistics

Exercise

1 A hockey team played 20 matches during a season. The numbers of goals they scored were as follows.

Number of goals	0	1	2	3	4	5	6	7	8	9
Frequency (number of matches)	2	7	4	3	2	1	0	0	0	1

a) Draw a vertical line chart to illustrate these data.

b) Find the mean, mode, median and range of the number of goals per match.

c) In one game the team scored 9 goals; the other team that day were all suffering from a late party the night before. If that match is ignored, what is the effect on each of the answers to part b)?

2 In a traffic survey, the number of people in each car passing a checkpoint was recorded. The results are shown in this bar chart.

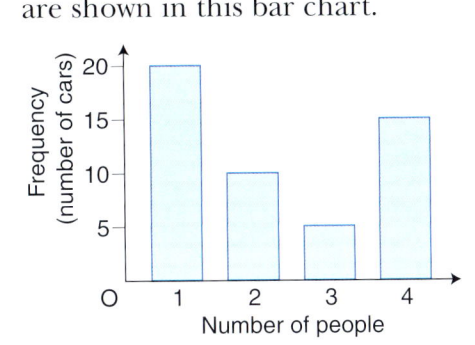

a) Make a frequency table for these data.
b) Describe the pattern of the data in words.
c) Find the mean, mode and median of the data.
d) Which of these could you find just by looking at the bar chart?
e) Why does none of these averages tell you the full picture?
f) Write down the mean number of *passengers* per car.

3 This pie chart shows the number of cars that Mr Smart sold per week for the last 60 weeks.

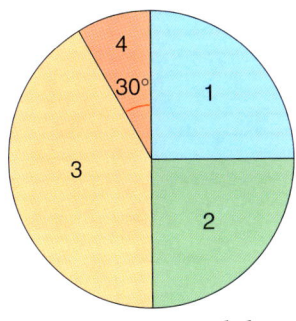

a) Write down the modal number of cars sold per week.
b) Write down the median.
c) Make a frequency table.
d) Calculate the mean.
e) Mr Smart is paid a bonus of £200 for each car he sells. Which figure does he most want to be high, the mode, mean or median?

Some calculators can work out the mean of data given in a frequency table; with others you have to use brackets or the memory.

Write down clear instructions telling someone how to use your calculator to find the mean of data given this way.

13: Practical statistics

Making comparisons

Many professional basketball players are very tall. Imran collects these data on the heights of 120 basketballers and 120 footballers (all to the nearest whole inch).

Height (inches)	70	71	72	73	74	75	76	77	78	79	80	81
Footballers	0	12	24	31	34	12	4	2	1	0	0	0
Basketballers	0	0	0	0	4	10	25	37	25	12	7	0

He draws these two frequency polygons showing the players' heights.

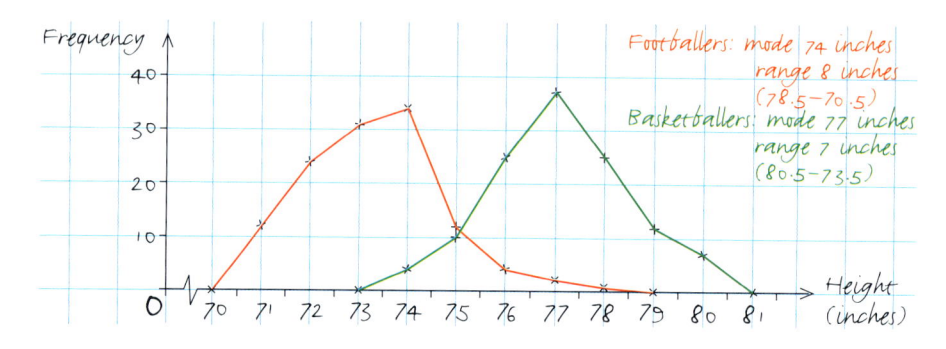

Footballers: mode 74 inches, range 8 inches (78.5 − 70.5)
Basketballers: mode 77 inches, range 7 inches (80.5 − 73.5)

? What does this graph tell you?

Imran decides to plot cumulative frequency curves for the same data.

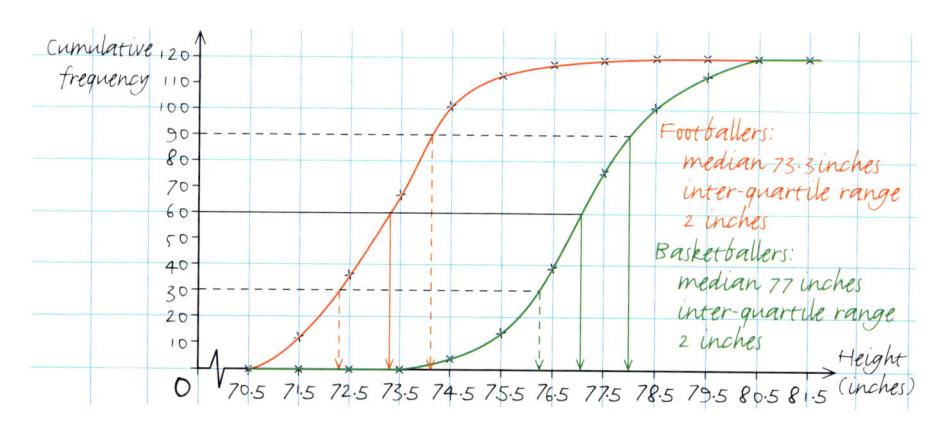

Footballers: median 73.3 inches, inter-quartile range 2 inches
Basketballers: median 77 inches, inter-quartile range 2 inches

? What does this graph tell you that the other one does not?

Which graph do you find more helpful for comparing the heights of footballers and basketball players?

13: Practical statistics

1 Emma has done a survey of the number of items in people's shopping baskets at the checkouts of two different shops. Here are her results.

Number of items	1-5	6-10	11-15	16-20	21-25	26-30	31-35
Frequency: shop A	24	49	13	8	4	2	0
Frequency: shop B	5	8	15	33	29	10	0

a) Which is the modal group for each shop?

b) On the same set of axes, draw the frequency polygon for each shop.

c) Compare the frequency polygons. What do they tell you about the two shops?

d) Suggest what type of shop each might be.

2 This table shows the results of a survey that was done to check the phasing of the traffic lights at the site of some major roadworks. The time spent waiting by 200 cars travelling in each direction was recorded one morning.

Delay, t (minutes)	$0 \leq t < 2$	$2 \leq t < 4$	$4 \leq t < 6$	$6 \leq t < 8$	$8 \leq t < 10$	$10 \leq t < 12$	$12 \leq t < 14$	$14 \leq t < 16$
North-bound	50	80	60	10	0	0	0	0
South-bound	10	25	35	55	45	20	10	0

a) On the same set of axes, draw a cumulative frequency curve for the delay times in each direction.

b) Estimate the median and the inter-quartile range for each direction.

c) Do you think the traffic lights are sensibly phased?

d) Might the survey have produced different results at a different time of the day?

Rewrite the height data you collected on page 111 in separate frequency tables for boys and girls. (Ideally you need the same number of boys and girls.) Use frequency polygons and cumulative frequency curves to compare the heights of the boys and the girls.

13: Practical statistics

Conducting a survey

A group of students have been asked to do some market research for a firm wanting to know the proportion of left-handed people in the population. They also want to know whether left-handedness is related to scientific, musical, artistic or sporting ability, and what tasks are particularly difficult. They will then have some idea of the potential market for gadgets designed for left-handers.

We need to do a survey Come on, let's go and ask people.

Not yet. We need to plan first.

Let's share the work. Each person can ask one registration group, then we can pool the answers.

Good idea, but in that case, we must all use the same questionnaire.

What do we want to know?

Male or female? It might make a difference.

Left or right-handed? Will people mind being asked that?

I don't think so. It isn't considered bad to be left-handed these days.

What subjects?

No, we don't need to ask that. It's on the college database. So is male or female. We only need to ask if they're left-handed.

When we know who the left-handers are, we'll run a separate survey of them to find out what gadgets they would find useful.

13: Practical statistics

1 Below is a list of important points to remember when you do a survey. Most of these were brought up by the students.

Who made each point?

Were all the points covered?

> **Guidelines for surveys**
>
> - Always plan before starting to collect lots of data.
>
> - Don't duplicate effort. Share **primary data** (the data you collect), or use **secondary data** (data that is already available) when you can.
>
> - Put questions in a logical order.
>
> - Keep questions simple, with definite answers.
>
>   A **closed question** has only a fixed number of possible answers (such as yes/no/don't know) from which the respondent is asked to select.
>
>   An **open question** has a wide range of possible answers.
>
>   Closed questions produce answers that are easier to process.
>
> - Be sensitive about asking embarrassing questions.
>
> - Check that you have asked everything you want to know.
>
> - Run a **pilot survey** to test your questionnaire (ask just a few people to answer it, and see if any questions cause difficulties).

Design a questionnaire about TV viewing habits.

2 Tim and Jeremy don't like the lunches served in the canteen. They decide to do a survey of customer opinion so that the canteen can provide what people want.

Tim and Jeremy each design a questionnaire for the survey. Tim includes some questions on other matters that interest him.

Compare the two surveys. List your criticisms of each one.

Discussion

Jeremy's questionnaire

Questionnaire
We are considering altering lunchtime catering arrangements. It would be helpful to know your view.

1. Please tick your preference
   - Cooked meal ☐   Light snack ☐
   - Salad bar ☐   Sandwiches ☐

2. Please tick if you have special dietary requirements
   - Vegetarian ☐   Halal meat ☐
   - Vegan ☐   Kosher ☐
   - Other ☐

3. Would you prefer
   - Coffee ☐   Tea ☐
   - Coke ☐   Lemonade ☐
   - Other _____

Thank you for completing my questionnaire. Please return to reception.

Tim's questionnaire

Questionnaire
1. Do you have a cooked meal or salad or sandwiches for lunch?
2. What do you think of our chances in the World Cup?
3. If you don't have any of them what do you have for lunch?
4. How much telly did you watch last year?
5. Aren't you glad we're going to replace yucky meals with some good nosh?

153

13: Practical statistics

Writing your report

When you have done a survey, the report you write about it should include the following sections.

Executive summary — This explains in a few lines the main points to emerge from your survey.

> These results show that "1 in 10" seems to be a very good "rule of thumb" for the population proportion. The only significant subject values are for English and Languages suggesting that there may be a connection between linguistic ability and left-handedness. We would suggest a possible focus on institutions specialising in languages; perhaps offering them resource packs containing advertisements and

Introduction — This explains what you wanted to find out, and why.

> "Southpaws" commissioned a survey into left-handedness and its associated patterns in order to pinpoint possible target areas for current and future sales campaigns. In particular the survey was asked to find out whether the quoted "1 in 10 of the population is left-handed" is correct; to determine possible patterns of interests, sex, and genetics, and to collect ideas for future use

Approach — This explains how you collected your data – how many people you asked, how you selected them, what you asked them (include a copy of your questionnaire).

> We decided to use members of the general public to establish the proportion of left-handers, and the patterns of incidence of left-handedness; and then question a sample of left-handers to establish potential market need. The first objectives, those of discovering the number (and names) of left-handers together with information as to left-handedness in other members of college families, were achieved by a

Findings — This should present the data in the most suitable ways. Look back at Chapter 7 to remind yourself of the types of diagram that are possible.

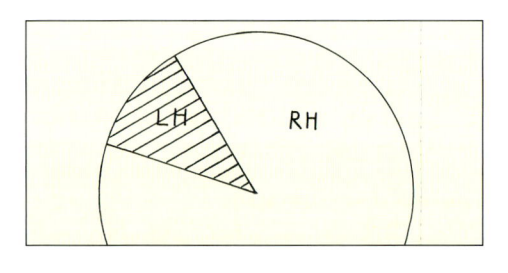

Conclusions and recommendations

> The survey suggests that there is a potential market for Southpaws products but it would need a very strong advertising campaign to make people aware of the possibilities. Our findings suggest that left-handedness is at least partly an inherited characteristic, that the "1 in 10" theory is generally true but that the proportion may be slightly larger. In our Sixth form college the proportion, while still within

13: Practical statistics

In order to conduct a survey and write a report you need to be able to

★ design a questionnaire

★ collect, collate and display data

★ use the data to draw conclusions

★ recognise when secondary data are available

★ present your findings clearly in a report.

What's the difference between primary and secondary data, then? I mean, which one is best for coursework?

Primary data is best. You collect it yourself so you get exactly what you want.

But you don't get very much. If you use secondary data you have a large number of observations and they have been collected properly.

I'm going to do a survey on a sample, that's primary, and then compare it with some secondary data. That way I'll have the best of both worlds.

What's the difference between a sample and a census, then?

A sample is a fairly small number of people but a census is a survey of everybody. Sampling is much cheaper but a census is more reliable.

They take a national census every 10 years. It covers everyone in the whole country. The information is confidential for 100 years. The 1881 census wasn't available until January 1982.

I asked my Mum what her Granny was called and used the 1881 census at the library. There is a surname index for the whole country, in alphabetical order, just like our registers. I just looked her up and found out where she was living, all her brothers and sisters, what her parents did, where they were born and how old they were. Fantastic!

Now you can do your own survey. It could be the one described below, or one on a topic of particular interest to you.

Here are some secondary data from *Social Trends 26* (1996).

Design your own questionnaire to collect matching primary data.

Test your questionnaire on 2 or 3 people you know. When you are satisfied that you have got it right, use it to do a survey of sporting activity in your school or college or local area.

Write a report on your findings.

Participation[1] in the most popular sports, games and physical activities: by gender and age, 1993-94

United Kingdom — Percentages

	16-19	20-24
Males		
Walking	45	46
Snooker/pool/billiards	56	47
Swimming	23	19
Cycling	37	19
Soccer	44	27
Golf	15	13
Females		
Walking	40	41
Keep fit/yoga	29	28
Swimming	26	25
Cycling	14	12
Snooker/pool/billiards	26	17
Tenpin bowls/skittles	9	9

1 Percentage in each age group participating in each activity in the four weeks before interview.

Source: General Household Survey, Office of Population Censuses and Surveys; Continuous Household Survey, Department of Finance and Personnel, Northern Ireland

Fourteen

Finding lengths

Before you start this chapter you should

★ be familiar with metric and Imperial units

★ know that the angles in a triangle add up to 180°.

Scale drawings

A chapter on bones in a biology textbook contains these pictures.

Human tooth

Scale 5:1 ← *Read this as 'five to one'*

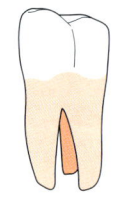

The tooth is small in real life, so an enlargement has been drawn. The scale of the drawing is 5:1. This means that 5 cm on the drawing represents 1 cm in real life. The drawing is 5 times as long as the actual tooth.

Human skull

Scale 1:10 ← *Read this as 'one to ten'*

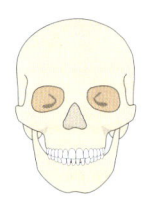

The skull is too big in real life to draw full size, so the drawing has been reduced. The scale is 1:10. This means that 1 cm on the drawing represents 10 cm in real life. The actual skull is 10 times as high as the drawing.

When you see a scale like 5:1 or 1:10, remember that the first number refers to the drawing and the second number refers to real life.

When making a scale drawing, you do calculations like these:

> Length of tooth in drawing = actual length × 5
> Height of skull in drawing = actual height ÷ 10

When reading or using a scale drawing, you do calculations like these:

> Length of tooth = length of drawing ÷ 5
> Height of skull = height of drawing × 10

If you are not sure whether to multiply or divide, think about whether you want to make the object bigger or smaller!

14: Finding lengths

Exercise

1 a) A swimming pool 50 m by 20 m, is drawn to a scale of 1:250. How big is it in the drawing?

b) A ladybird, length 8 mm and height 5 mm, is drawn to a scale of 5:1. How big is it in the drawing?

2 This plan of Laura's bedroom is not drawn to scale.

A is the bed
(180 cm by 80 cm)

B is the wardrobe
(90 cm by 50 cm)

C is the dressing table
(100 cm by 40 cm)

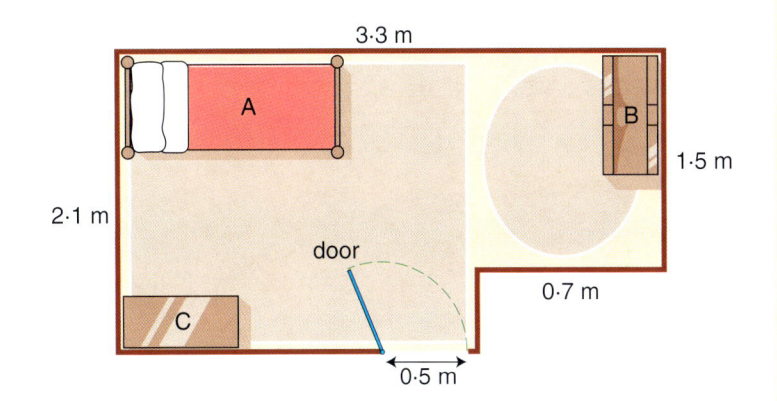

a) Draw an accurate plan of Laura's bedroom using a scale of 1:20.

b) Measure the diagonal distance across the drawing from the corner with the dressing table to the corner with the wardrobe.

Work out what this distance would be in real life.

c) Use a compass to draw the arc through which the end of the door travels when it is opened or closed.

Measure how close this arc goes go the edge of the bed.

What would this distance be in real life?

Is this enough space to walk through?

3 a) This diagram of a red blood cell has a diameter of 3 cm.

What is the diameter of the real red blood cell?

b) A real human hair has a width of 0.1 mm.

If you drew it to the same scale as the red blood cell, what would be the width of the hair in the picture?

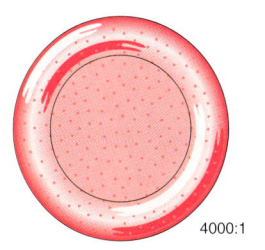

It is 140 km from Swansea to Plymouth, as the crow flies. Find a map with both places on the same page. Measure the distance between them in centimetres.

Use your answer to work out the scale of the map.

14: Finding lengths

Using bearings

An aeroplane is flying from London to Manchester.

The pilot needs to know exactly what direction to take.

She can do this by measuring the angle on the map between the direction she needs to fly in and a line going north from London.

The angle is measured clockwise from the North line.

Use an angle measurer to measure this angle.

You should find that it is 330 degrees.

This angle is called the **bearing** of Manchester from London.

Remember that bearings are always measured clockwise from north.

Bearings are always written with three figures.

A bearing of 62° is written as 062°.

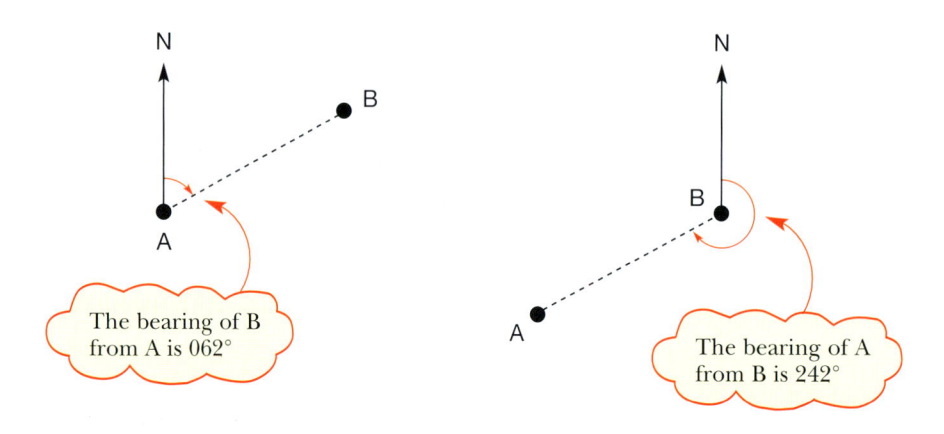

The bearing of B from A is 062°

The bearing of A from B is 242°

14: Finding lengths

1 A, B, C and D are 4 ships.

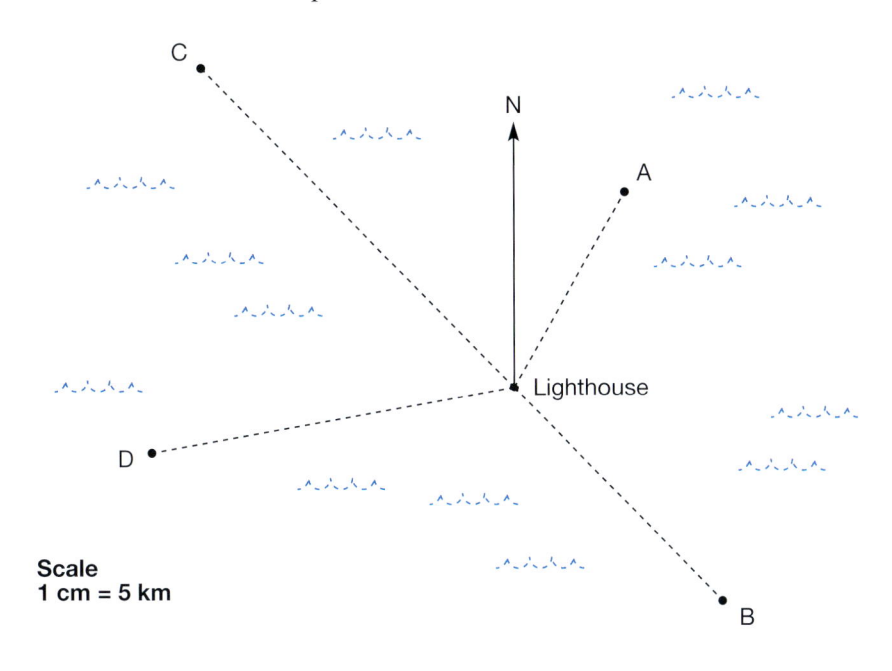

Scale 1 cm = 5 km

a) Measure the bearing of each of the ships from the lighthouse.

b) Find the distance of each ship from the lighthouse.

2 This diagram shows three towns, A, B and C.

Measure the bearing of

a) A from B   b) B from A

c) A from C   d) C from A

e) B from C   f) C from B

g) What do you notice about each pair of bearings?

3 Andrew and Ranjit are on a hike in the hills.

They find that a nearby hill, Mill Crag, is on a bearing of 158°.

They walk for 4 km north-east and then find that Mill Crag is on a bearing of 205°. Make a scale drawing of their journey, using a scale of 1:50 000.

Use your drawing to find out how far they are from Mill Crag at each of the points where they take a bearing.

Plan a walk through open country.

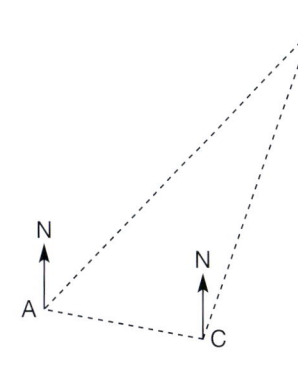

14: Finding lengths

Pythagoras' rule

Robert is a farmer and is building a gate for one of his fields. He sketches out a rough design for the gate so that he can work out what lengths of wood he needs.

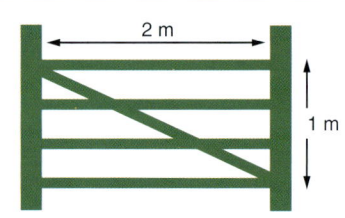

It is easy for Robert to cut the lengths of wood he needs for the horizontal bars, but he does not know how long the diagonal piece needs to be.

He could make a scale drawing and measure the length of the diagonal piece, but it would be much easier if there was a simple way to calculate the length. Luckily, there is!

? *Measure these triangles and find the areas of squares A, B and C in each diagram. What do you notice?*

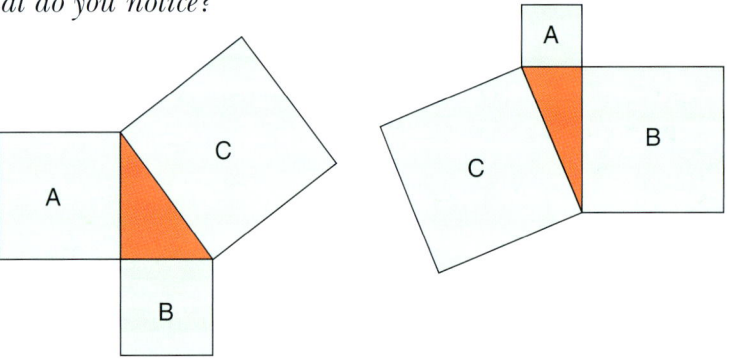

You should have found that the areas of the two smaller squares add up to the area of the largest square in each case.

This rule is called Pythagoras' rule (or theorem). It is true for all right-angled triangles. It is usually written like this:

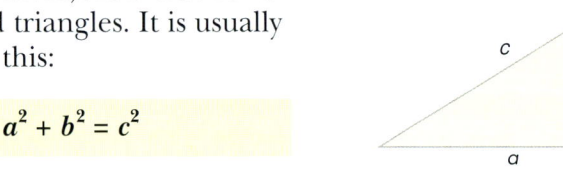

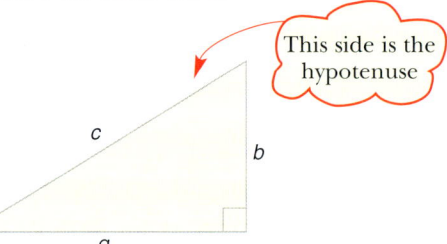

$$a^2 + b^2 = c^2$$

This side is the hypotenuse

The side labelled c must be the hypotenuse (the longest side, always the one opposite the right angle).

The diagonal bar on Robert's gate can be drawn as the hypotenuse of a right-angled triangle like this:

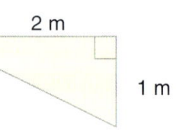

Pythagoras' rule is: $a^2 + b^2 = c^2$

In this case: $2^2 + 1^2 = c^2$

$4 + 1 = c^2$

$5 = c^2$

$c = \sqrt{5} = 2.236$

*$\sqrt{5}$ is called a **surd**. It does not have an exact numerical value*

'Undo' the square by finding the square root

So the diagonal piece on Robert's gate needs to be 2.24 m long.

14: Finding lengths

1 Use Pythagoras' rule to find the length of the hypotenuse in the triangles below.

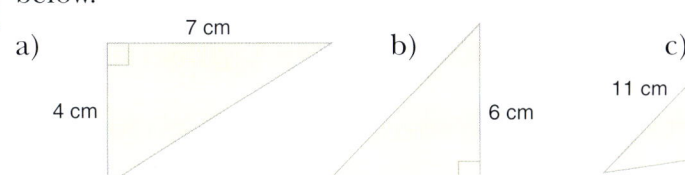

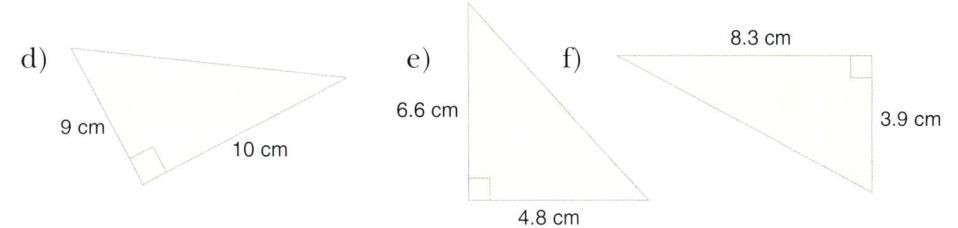

a) 7 cm, 4 cm
b) 6 cm, 6 cm
c) 11 cm, 8 cm
d) 9 cm, 10 cm
e) 6.6 cm, 4.8 cm
f) 8.3 cm, 3.9 cm

2 A field is 150 metres long and 120 metres wide. A footpath goes diagonally across the field. How long is the footpath?

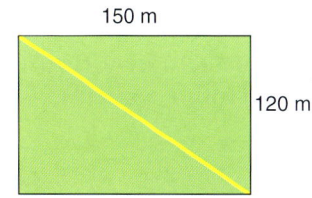

3 A ship sails 23 km due north and then 17 km due east. It then sails back to its starting point in a straight line. How far is the distance back to the starting point?

4 The diagram shows the two points A (1, 2) and B (4, 4).

a) What is the horizontal distance from A to the point P?

b) What is the vertical distance from B to the point P?

c) Use Pythagoras' rule to find the distance from A to B.

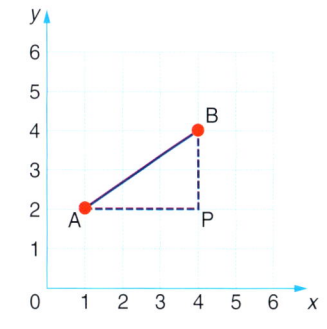

5 Use the same method as in question 4 to find the distances between each of the following pairs of points, leaving your answers as surds.

(You will find it helpful to draw diagrams showing the points.)

a) (3, 1) and (5, 6)   b) (4, 2) and (1, 5)
c) (−2, 4) and (3, 5)   d) (0, 3) and (−2, −1)

161

14: Finding lengths

Finding one of the shorter sides

Terry is a window-cleaner. His ladder is 8 metres long.

For safety reasons he always places the foot of the ladder at least 1.5 metres from the wall. He wants to know how far up the wall he can make his ladder reach.

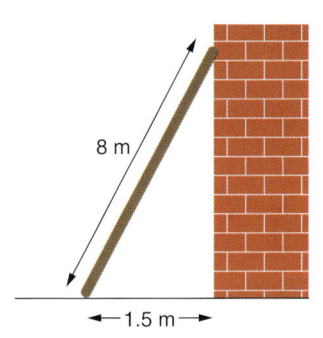

So far you have only been asked to find the length of the hypotenuse in a right-angled triangle. To solve Terry's problem, you need to be able to find one of the two shorter sides. You can use Pythagoras' rule to solve this kind of problem as well.

This is a simplified diagram of Terry's ladder.

y stands for the height up the wall that the ladder reaches.

Pythagoras' rule is

$$a^2 + b^2 = c^2$$

In this case

$$1.5^2 + y^2 = 8^2$$
$$2.25 + y^2 = 64$$

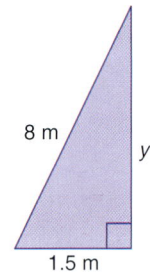

To find y^2, you need to subtract 2.25 from both sides of the equation.

$$y^2 = 61.75$$

Now you can find y by taking the square root of 61.75.

$$y = \sqrt{61.75} = 7.86$$

The ladder reaches 7.86 metres up the wall.

Remember!

To find the hypotenuse, you have to **add**.

To find one of the shorter sides, you have to **subtract**.

14: Finding lengths

1 Find the lengths of the sides marked *x* in each of these triangles. In some of them you have to find the hypotenuse, in others you have to find one of the shorter sides. Leave *x* as a surd in a), b) c) and f).

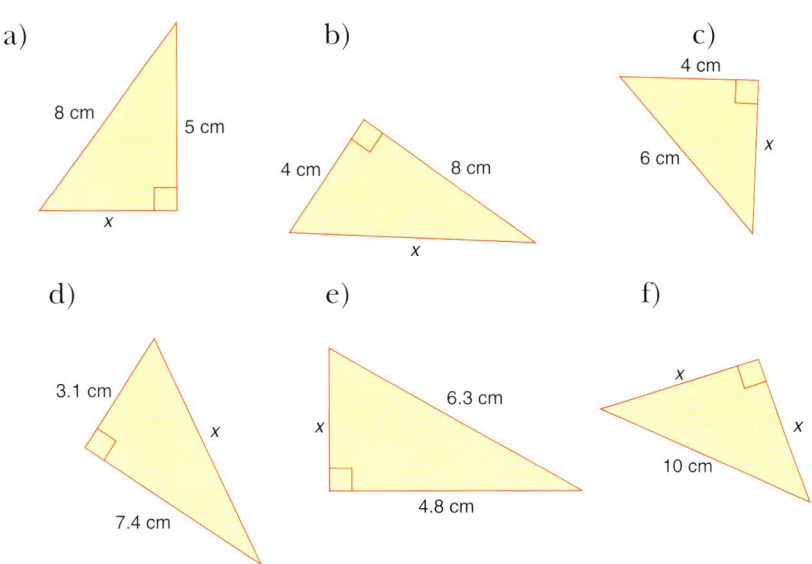

a) 8 cm, 5 cm, x

b) 4 cm, 8 cm, x

c) 4 cm, 6 cm, x

d) 3.1 cm, 7.4 cm, x

e) 6.3 cm, 4.8 cm, x

f) x, 10 cm, x

2 A ladder 6.2 metres long is to be placed so that it just reaches a window 5.7 metres from the ground. How far from the wall is the foot of the ladder?

3 The diagram shows an isosceles triangle split into two congruent right-angled triangles.

a) Use Pythagoras' rule to find the height, *y*, of the triangle, leaving *y* as a surd.

b) Find the area of the triangle in surd form.

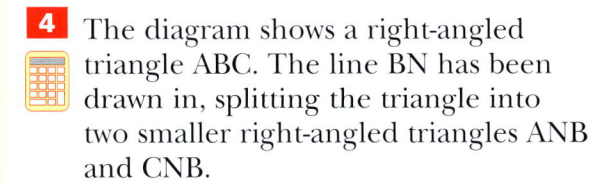

(6 cm, 6 cm, y, 4 cm)

4 The diagram shows a right-angled triangle ABC. The line BN has been drawn in, splitting the triangle into two smaller right-angled triangles ANB and CNB.

(AB = 4 cm, AC = 5 cm)

a) Work out the length of the side BC.

b) Using AB as the base of the triangle, work out the area of the triangle.

c) Using AC as the base of the triangle, use your answer to b) to work out the length of BN.

d) Work out the lengths of AN and CN.

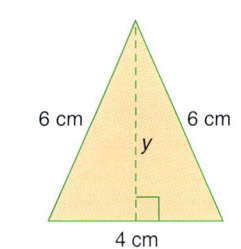

163

14: Finding lengths

Finishing off

Now that you have finished this chapter you should know how to

★ make and interpret scale drawings

★ use 3-figure bearings

★ use Pythagoras' rule to find the unknown side of a right-angled triangle, given the other two sides.

Use the questions in the next exercise to check that you understand everything.

Mixed exercise

1 Karen is writing a book on insects. She wants to include a picture of a flea that is 0.3 cm long in real life. She needs to decide what scale to use for the drawing.

Find the length of the flea in the drawing if Karen uses a scale of

a) 20:1

b) 1:5

c) 500:1

Which scale do you think would be the most suitable?

2 Here is a rough plan of a one-bedroomed flat.

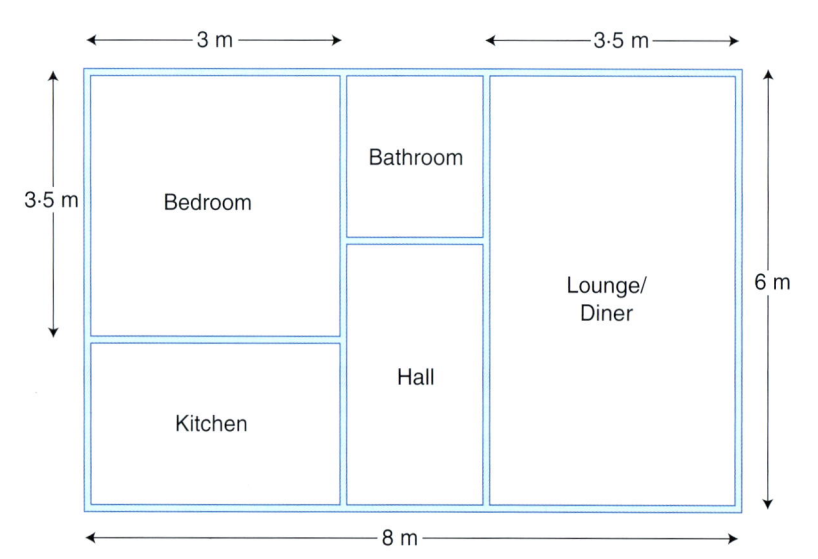

a) Make a scale drawing of the flat, using a scale of 1:50.

b) Each door in the flat is 75 cm wide. Add doors to your scale drawing in suitable places.

164

14: Finding lengths

Mixed exercise

3 Find the lengths of the sides marked with letters in these triangles.

a) [triangle: 11 cm, 15 cm, a, right angle]

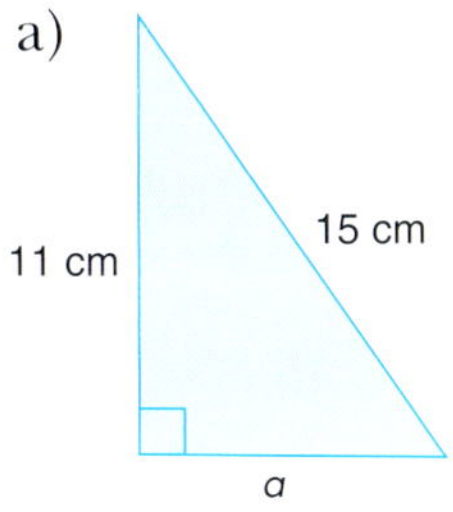

b) [triangle: 8 cm, 5.5 cm, b, right angle]

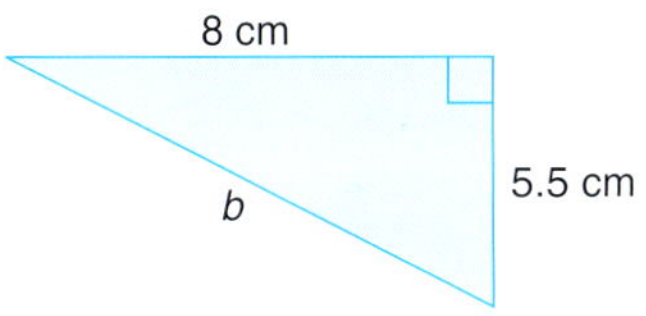

c) [triangle: c, c, 9 cm, right angle at top]

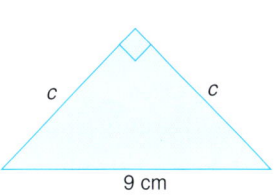

d) [triangle: 8.9 cm, 4.6 cm, d, right angle]

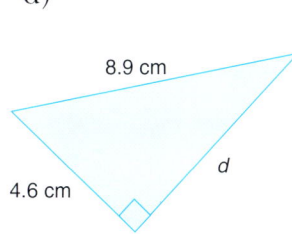

e) [triangle: e, 7.7 cm, 10.1 cm, right angle]

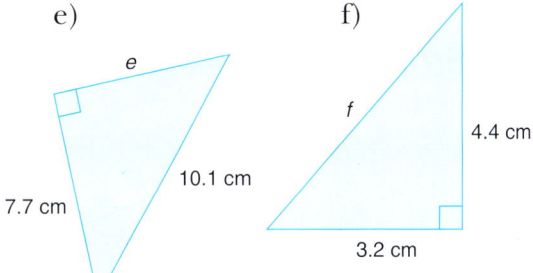

f) [triangle: f, 4.4 cm, 3.2 cm, right angle]

4 Find the distance between each pair of points.

a) (1, 4) and (4, 0)

b) (−2, 3) and (2, −1)

c) (−3, −4) and (−1, 1)

5 A ship leaves the port of Harwich (on the east coast) and sails 30 km. It is then 12 km north of Harwich.

How far east is it from Harwich?

Investigation

This triangle is a right-angled triangle.

$3^2 + 4^2 = 9 + 16 = 25 = 5^2$

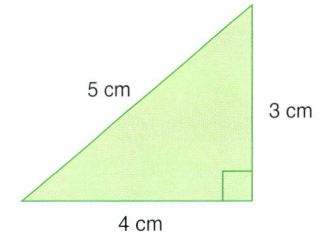

The numbers 3, 4, 5 are called a **Pythagorean triple** because they obey Pythagoras' rule.

Find as many different Pythagorean triples as you can.

(Don't count triples which are just a multiple of one you have already found, like 6, 8, 10 or 9, 12, 15, which are both multiples of 3, 4, 5.)

[right-angled triangle: 5 cm, 3 cm, 4 cm]

Fifteen

Money

Simple interest

Katie has saved £3000 and wants to invest it for two years in a building society. These are the rates on offer.

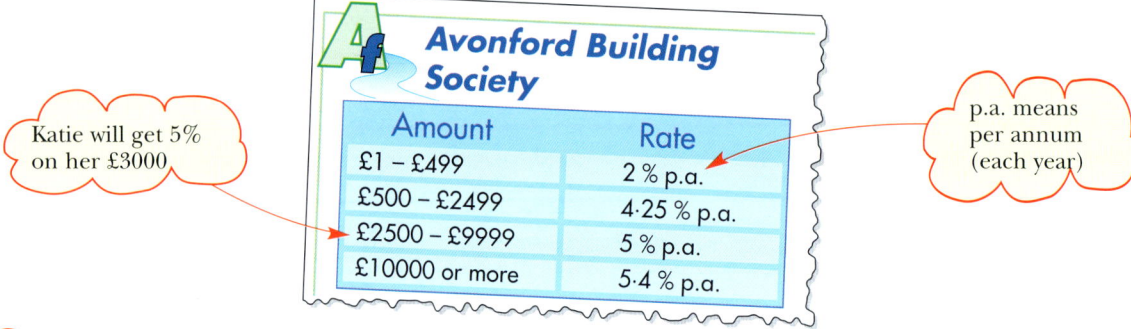

Avonford Building Society

Amount	Rate
£1 – £499	2 % p.a.
£500 – £2499	4·25 % p.a.
£2500 – £9999	5 % p.a.
£10000 or more	5·4 % p.a.

Katie will get 5% on her £3000

p.a. means per annum (each year)

? *Why do you think the interest rate is higher on larger amounts?*

Katie works out her yearly interest like this:

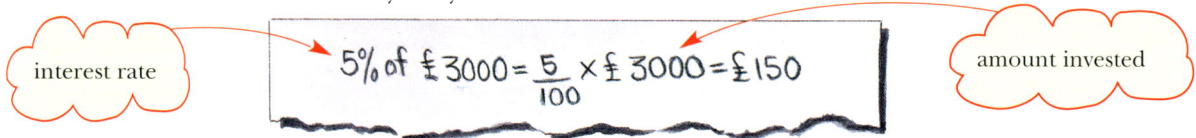

$$5\% \text{ of } £3000 = \frac{5}{100} \times £3000 = £150$$

interest rate — *amount invested*

The £150 interest is sent to Katie at the end of the first year.

The £3000 in the account will earn another £150 interest during the second year.

The **simple interest** Katie earns over 2 years will be

$$2 \times £150 = £300$$

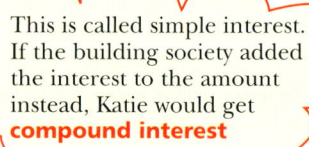

*This is called simple interest. If the building society added the interest to the amount instead, Katie would get **compound interest**.*

? *How much simple interest would Katie earn over 5 years?*

You can use this formula for calculating the simple interest, I:

$$I = \frac{P \times R \times T}{100} = \frac{PRT}{100}$$

where P is the money invested (sometimes called the principal),

R is the rate in % p.a.,

T is the time in years.

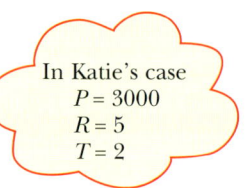

In Katie's case
$P = 3000$
$R = 5$
$T = 2$

? *Check that this formula gives £300 for Katie's simple interest over 2 years. Explain why this formula works.*

15: Money

1 Calculate the simple interest on

a) £600 invested at 5% p.a. for 1 year;

b) £3000 invested at 5.5% p.a. for 2 years;

c) £1240 invested at 4.75% p.a. for 4 years;

d) £5000 invested at $6\frac{1}{2}$% p.a. for 3 years;

e) £850 invested at 4% p.a. for 6 months;

f) £50 000 invested at 6.85% p.a. for 2 years.

2 The members of the Lewis family want to maximise their investment income. Grant has £700 to invest, Maxine has £6500 and Scott has £1250.

They have collected these leaflets.

Southern Building Society 4·5% Minimum £500

Western Building Society 5% Minimum £1000

Eastern Building Society 5·5% Minimum £5000

a) Say where each person should invest, and what simple interest he or she will receive over 3 years.

b) How much interest would Scott lose by investing in the Southern Building Society instead?

3 The Mehta family has this leaflet from Ashtown Building Society.

How much simple interest will be earned by

a) Mohan investing £1500 for 3 years?

b) Sunil investing £5000 for 2 years?

c) Alka investing £750 for 3 years?

Ashtown Building Society

Amount	Rate of Interest
up to £2000	3·75 %
£2000–£4999	4·50 %
£5000 or more	4·75 %

Mohan and Alka decide to invest all their money in a joint account.

They share the interest in proportion to the money they invested.

d) What rate of interest will they earn now?

e) How much extra interest does each get over the 3-year period?

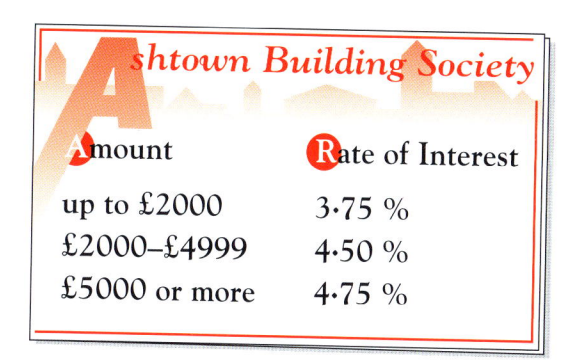

Go to 4 building societies or banks and find out the rates of interest they offer on 'instant access' accounts.

What is the highest rate you can get on a) £200? b) £1500? c) £6000?

15: Money

Compound interest

George puts £400 in this building society.

This stands for 'per annum' or per year

BUILDING SOCIETY
TopHat
Platinum Savings Account
10% p.a. interest

How much money does he have after 3 years?

Building societies and banks usually pay interest straight into your account.

At the end of the first year, George gets the interest on his £400.

At the end of the second year, he gets the interest on his £400 and on the interest he got last year. This is the **compound** interest.

This is how George's money increases over 3 years.

After 1 year:

   Interest = $\frac{10}{100}$ × £400 = £40

   Amount at the end of year = £400 + £40 = £440

After 2 years:

   Interest = $\frac{10}{100}$ × £440 = £44

   Amount at end of year = £440 + £44 = £484

After 3 years:

   Interest = $\frac{10}{100}$ × £484 = £48.40

   Amount at end of year = £484 + £48.40 = £532.40

After 3 years, George has £532.40 in his account.

After n years the amount in George's account is 1.1^n. Explain this formula.
What is the general formula for the amount that £P becomes after being invested for n years of R% compound interest?

How much would George have if the building society paid simple interest (interest just on the original £400)?

Would you prefer simple or compound interest?

15: Money

1 Calculate the compound interest on

a) £200 invested at 10% p.a. for 3 years

b) £1000 invested at 5% p.a. for 2 years

c) £150 invested at 6% p.a. for 2 years

d) £14 500 invested at 9% p.a. for 3 years

e) £22 420 invested at 8% p.a. for 4 years

f) £6200 invested at 7.5% p.a. for 2 years.

2 Jan invests £1500 in a bank account that pays interest at 6% p.a.

How much does she have after 3 years if the bank pays

a) simple interest?

b) compound interest?

3 Andy invests £800 in a building society account that pays compound interest at 8% p.a.

Copy and complete this chart which shows details of his account over 5 years.

Year	Amount at start of year	Interest	Amount at end of year
1	£800	£800 × $\frac{8}{100}$ = £64	£800 + £64 = £864
2	£864	£864 × $\frac{8}{100}$ =	
3			
4			
5			

4 Tessa has £1200 to invest. She sees these adverts.

How much does Tessa have after 2 years if she invests in

a) the bank?

b) the building society?

c) Tessa decides to invest in the bank. How long does she have to leave the money in this account before this becomes a bad decision?

Gray Bank — simple interest 7% p.a.

Austin and Pearson Building Society — 6.5% p.a. COMPOUND INTEREST

169

15: Money

Bills

Emily moved into a bedsit on 21 July. The electricity meter readings when she moved in and again 3 months later are shown below.

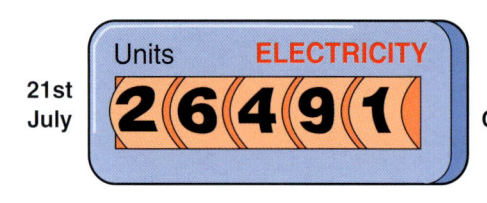

21st July — Units ELECTRICITY **26491**

21st October — Units ELECTRICITY **26933**

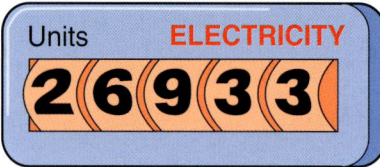

This is the bill that Emily received.

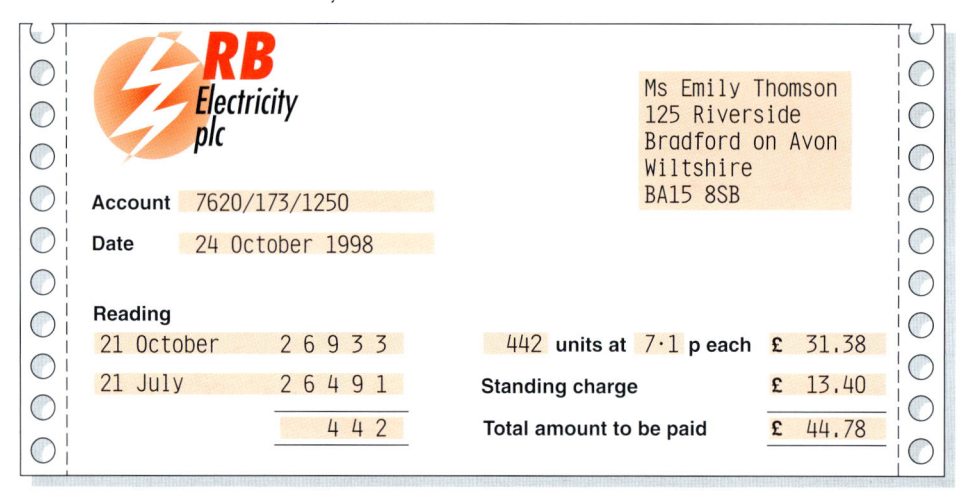

RB Electricity plc

Ms Emily Thomson
125 Riverside
Bradford on Avon
Wiltshire
BA15 8SB

Account   7620/173/1250
Date   24 October 1998

Reading
21 October   2 6 9 3 3
21 July   2 6 4 9 1
              4 4 2

442 units at 7·1 p each   £ 31.38
Standing charge   £ 13.40
Total amount to be paid   £ 44.78

How many units has Emily used?

How is this worked out?

How much does one unit of electricity cost?

The number of units Emily has used is the difference between the two readings.

Emily has used 442 units. The cost of 1 unit is 7.1 pence.

You work out the total cost of units like this:

   442 × 7.1p = 3138.2p

This is 3138p to the nearest penny

What is this in pounds and pence?

The standing charge is the same every quarter: it is £13.40 however much electricity Emily uses.

The standing charge is added to the cost of the units to work out the total amount to be paid.

Do you think that Emily will use the same amount of electricity in the next three months? Explain your answer.

15: Money

1 Here are two of Meryl's gas meter readings.

29th October: 11439
29th January: 12044

a) How many units has Meryl used in this quarter?

b) Each unit costs 13.2p. How much will the units cost in total?

c) Meryl pays a standing charge of £14.80 a quarter. Work out her total bill.

2 On 2 April the reading on Stephen's water meter was 1320.
On 2 October it was 1402.

a) How many units had he used in this time?

b) Each unit costs 54.5 pence. How much will they cost in total?

c) Stephen pays a half-yearly standing charge of £14.20. Work out his total bill.

3 Here are two of Kerry's electricity meter readings.

5th November: 7463
5th February: 8127

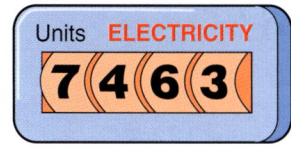

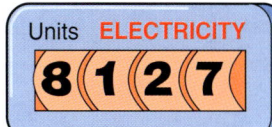

a) How many units has Kerry used in this quarter?

b) Each unit costs 7.6p. How much will these units cost in total?

c) Kerry pays a quarterly standing charge of £12.50. Work out her total bill.

d) For the next bill, due in May, Kerry has set aside £50.50.

How many units does she expect to use?

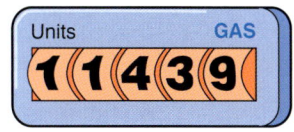

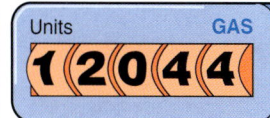

4 On 5 November Theresa's gas meter reading was 22836. On 5 February it is 23514. Units of gas cost 12.5 pence each. Theresa pays a quarterly standing charge of £11.75.

a) Work out her total quarterly bill.

b) From 5 February a new tariff is introduced. The cost of a unit of gas is increased by 0.1 pence but the standing charge is reduced by £4.20. Theresa uses the same amount of gas.

Is Theresa better off under the new tariff? Explain your answer.

c) What type of user would be better off on the new tariff?

Find six appliances that use electricity.

How much does it cost to use each one for an hour?

15: Money

Hidden extras

Buy now – pay later

Poppy wants to buy this television but she does not have £475.

She works out how much it will cost to buy it over 12 months.

TV £475 or 12 month terms

£475

Deposit   £100
12 payments of £40   £480
Total   £580

? *How much more is this than the cash price?*

Caution: Poppy must be very careful about doing a deal like this. If she fails to make the monthly payments, she may lose both the television and the money she has already paid.

Value added tax (VAT)

VAT is a tax which you pay when you buy either goods or services.

? *What is the present rate of VAT?*

Steve sees the same fax machine in two different stores.

Bill's Bargains £210 EX VAT

CERI'S CORNER £249 INCL VAT

Ex VAT means 'Excluding VAT'. The full price is £210 + VAT

Incl VAT means 'Including VAT'. The VAT has already been added on: £249 is the full price

At Bill's Bargains the VAT to be added is 17.5% of £210.

$$17.5\% \text{ of } 210 = \frac{17.5}{100} \times 210 = 36.75$$

The VAT is £36.75.

The full price is £210 + £36.75 = £246.75.

? *Which store is offering the better deal?*

Write down two reasons why Steve might still decide to buy the fax machine from Ceri's Corner.

? *Why do some stores display prices 'ex VAT'?*

15: Money

1 Work out the cost of each item spread over 36 months and the extra amount paid.

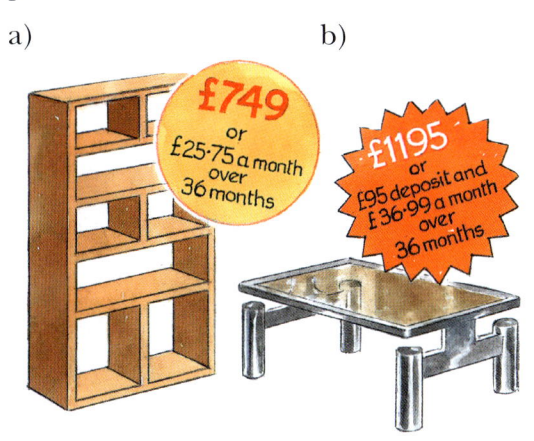

a) £749 or £25·75 a month over 36 months

b) £1195 or £95 deposit and £36·99 a month over 36 months

2 This table gives the repayments for each £100 of a loan.

Time in months	12	18	24	36
Repayment per month	£11.00	£8.10	£6.60	£5.00

How much is repaid in total on a loan of

a) £600 over 24 months?
b) £800 over 36 months?
c) £250 over 12 months?
d) £375 over 18 months?

3 The price of each of these items is given excluding VAT.

Work out the price of each item including VAT at 17.5%.

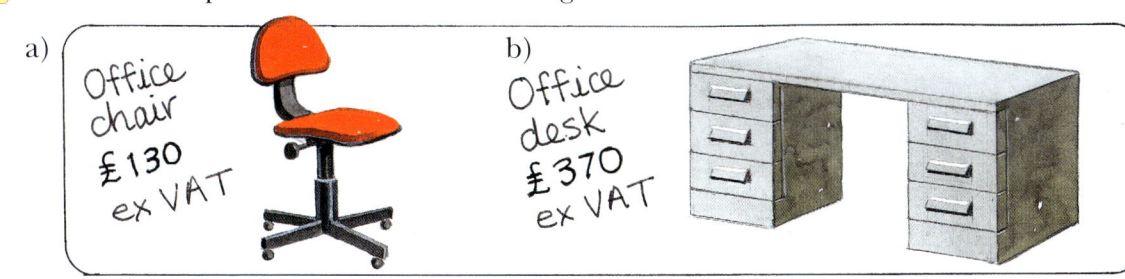

a) Office chair £130 ex VAT

b) Office desk £370 ex VAT

4 Store A £699, £119. Store B £149, £799.

Store A's prices exclude VAT at 17.5%. Store B's prices include VAT.

Which store offers the better deal and by how much for

a) the filing cabinet?
b) the computer?

a) The addition of VAT at 17.5% increases the price by about one sixth. Use this approximation to make a rough check of your answers to question 3.

b) You can work out VAT at 17.5% accurately without a calculator.
First you find 10% of the ex VAT price and write it down.
Then you halve it and write the new figure underneath, then you halve it again and write this new figure underneath.
Add up these three figures to get the amount of VAT.

Try this out on the items in question 3.
Why does it work?

15: Money

Profit and loss

Andy owns this sports clothing shop.

He buys the golf sweaters at £50 each and the ski-suits at £200 each.

Which item is more profitable for him?

At first glance the ski-suits look more profitable: Andy makes £20 profit on each suit, and only £10 on each jumper. But businesses are interested in profit as a percentage of cost:

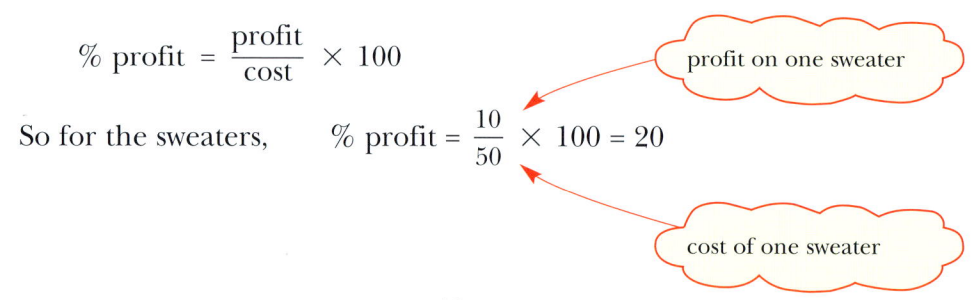

$$\% \text{ profit} = \frac{\text{profit}}{\text{cost}} \times 100$$

So for the sweaters, $\% \text{ profit} = \frac{10}{50} \times 100 = 20$

(profit on one sweater)
(cost of one sweater)

For the ski-suits, $\% \text{ profit} = \frac{20}{200} \times 100 = 10$

? *Why do businesses need to be concerned about percentage profit?*

? *Do you think Andy should stop selling ski-suits because they are less profitable?*

Andy does not always make a profit. He sells the last sweater for £48 in a sale. It cost him £50 so he makes a loss of £2.

? *Work out this loss as a percentage of the cost price.*

Andy buys track suits for £40 and prices them so as to make a 15% profit.

He works out the selling price like this:

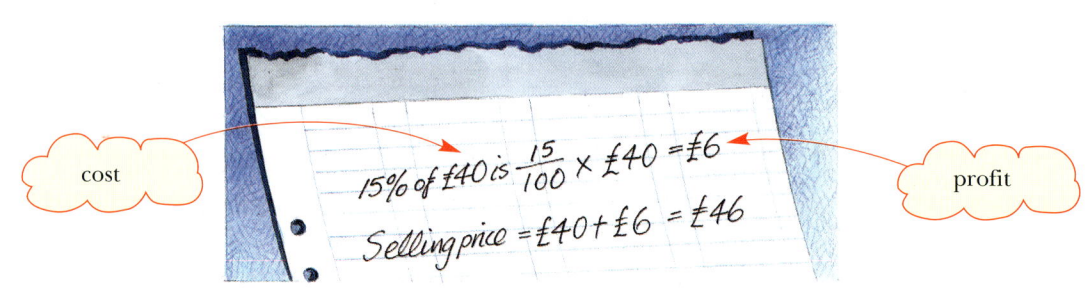

cost → 15% of £40 is $\frac{15}{100} \times £40 = £6$ ← profit

Selling price = £40 + £6 = £46

Note that the percentage profit is a percentage of the *cost price*, not of the *selling price*.

174

15: Money

1 Calculate the percentage profit or loss on

a) a table bought for £250 and sold for £380;

b) a coat bought for £80 and sold for £110;

c) a brief-case bought for £25 and sold for £17.50;

d) a tent bought for £35 and sold for £29.95.

2 Work out the selling price of

a) a clock bought for £60 and sold at a 20% profit;

b) a lamp bought for £62 and sold at a 40% profit;

c) a roof-rack bought for £80 and sold at a 15% loss;

d) a CD bought for £15 and sold at a 5% loss.

3 Carl buys 20 items for £600. He sells them at £39 each. Calculate

a) his total profit on the deal;

b) his percentage profit on the deal.

4 Tony buys jackets for £50 each and trousers for £20 each.

a) How much profit does he make on each item when he sells them in the sale?

b) Work out each profit as a percentage of the cost price.

5 Esme agrees to landscape a garden for £1000. She pays 4 other people to do the job for her.

She pays Mark £160 to lay a path.

She pays George £330 to install a pond and a fountain.

She pays Della and Kirsty £90 each to plant shrubs and bulbs.

a) How much money does Esme make on the deal?

b) Express your answer to a) as a percentage of Esme's outlay.

6 Chloe sells hats. She buys 12 of them for £7.50 each. She prices them so as to make a 40% profit and sells 10 of them at this price. The remaining 2 are sold off at £5 each. Calculate

a) her total receipts from the sales

b) her total profit

c) her percentage profit.

Find the share prices of 5 major companies 6 months ago and work out the percentage profit/loss a shareholder would have made by buying then and selling today.

15: Money

Repeated changes

Kelly works for Marina Drive car insurers.

She works out how much each driver has to pay, and then decides whether to give a discount by looking at this table.

MARINA DRIVE CAR INSURANCE

1. Apply no claims discount
2. Apply age discount
3. Apply post-code discount

No claims	Age	Post-code
1 year 10%	17–20 none	AB51 10%
2 years 25%	21–25 5%	AB52 10%
3 years 40%	26–34 10%	AB53 5%
4 years 60%	35+ 20%	AB54 none

Stewart's car insurance premium is £800 before discount.

He has 2 years no claims, is 23 and lives in AB51.

How much no claims discount does he get?

£800 is reduced by 25% for no claims.

$$25\% \text{ of } 800 = \frac{25}{100} \times \frac{800}{1} = 200$$

After his no claims discount Stewart's insurance is

£800 − £200 = £600

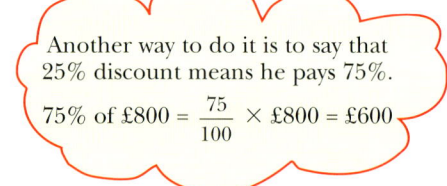

Another way to do it is to say that 25% discount means he pays 75%.

$$75\% \text{ of } £800 = \frac{75}{100} \times £800 = £600$$

How much age discount does he get?

£600 is reduced by 5% for age.

$$5\% \text{ of } 600 = \frac{5}{100} \times \frac{600}{1} = 30$$

After Stewart's age discount, his insurance is

£600 − £30 = £570

Another way to do it is to say that 5% discount means he pays 95%.

$$95\% \text{ of } £600 = \frac{95}{100} \times £600 = £570$$

How much post-code discount does he get?

How much does Stewart pay for his car insurance?

15: Money

1 Use the table on the opposite page to work out the cost of car insurance for each of these people.

   a) Callum's insurance before discounts is £900.
      He has 4 years no claims, is 49 and lives in AB54.

   b) Nathan's insurance before discounts is £690.
      He has 1 years no claims, is 19 and lives in AB52.

   c) Dorothy's insurance before discounts is £750.
      She has 3 years no claims, is 34 and lives in AB53.

2 Russell and Kay run a fitness centre. Last year, 3250 people a month visited the centre.

Russell thinks these figures will increase by 5% this year and 9% next year.

Kay thinks there will be a 7% increase this year and another 7% increase next year.

   a) How many people does Russell expect next year?

   b) How many people does Kay expect next year?

3 Oliver sells 350 CDs in October. He expects sales will go up by a fifth in November and by a third in December.

   a) How many CDs does Oliver expect to sell in November?

   b) How many CDs does Oliver expect to sell in December?

4 Natasha buys a second hand car for £4000. She estimates that each year its value will fall by 20% of its value at the start of the year.

   a) How much is it worth after one year?

   b) How much is it worth after two years?

   c) What percentage decrease in value has taken place over 2 years?

5 Bob runs a travel company. He expects to increase prices by 5% next year.

   a) How much will the cruise cost next year?

   b) He expects to increase prices the following year by 4%.

      How much will the cruise cost the following year?

   c) How would the answer to b) be affected if the prices were increased by 4% next year and 5% the following year?

Mediterranean cruise £1500

15: Money

Finishing off

Now that you have finished this chapter you should be able to

★ work out simple and compound interest
★ work out repeated proportional changes
★ work out household bills
★ work out the costs of 'buy now – pay later' deals
★ work out VAT
★ calculate profit or loss.

Use the questions in the next exercise to check that you understand everything.

Mixed exercise

1 Judy invests £4000 in a building society for 2 years at 5.5% p.a.
   a) How much simple interest does she earn?
   b) How much extra interest would she receive by investing in a building society offering her 5.75% p.a.?

2 Nina and Salil get these jobs.
   a) How much does Nina earn for a 37-hour week?
   b) How much does Salil earn for a 41-hour week?
   c) How many hours does Nina work in her second week to earn £266?

   Positions vacant – enquire within
   2 Store Assistants required
   £5.60 an hour
   37 hour week
   Overtime available at 'time and a half'

3 Each day 4500 cars and 325 lorries cross this bridge.

   BRIDGE TOLLS
   £1 per car
   £8 per lorry

   a) How much is paid in tolls each day?
   b) The car toll price is now increased by 30% and the lorry toll by 25%.
      The number of cars crossing decreases by 5% and the number of lorries decreases by 8%.
      How much is paid in tolls every day now?

4 Jo sells 80 tennis rackets in May. She expects sales to increase by 20% in June and fall by a third in July. How many tennis rackets does she expect to sell in July?

5 Joshua and Victoria borrow £60 000 to buy a house. The repayments are £475 a month over a period of 25 years.
   How much will they repay in total over this period?

6 Calculate the compound interest on
   a) £600 invested at 7% p.a. for 2 years
   b) £20 000 invested at 12% for 5 years
   c) £1250 invested at 8% p.a. for 3 years
   d) £6050 invested at 5.5% p.a. for 4 years.

15: Money

7 Joe invests £500 in Buckton Building Society.

At the same time, Mary invests £500 in Woodbridge Building Society.

Buckton Building Society — Super Saver Account — Simple interest 9% p.a.

Woodbridge Building Society — High Earners Account — 8% p.a. compound interest

a) After 2 years, how much interest has
   (i) Joe earned?
   (ii) Mary earned?

b) How many years is it before Mary has more in her account than Joe?

8 Tammy's gas meter reading was 4539 on 19 September and 4701 on 19 December. Units cost 48.2p and she pays a standing charge of £13.70 a quarter.

Work out her bill for this quarter.

9 Sal buys this settee over 30 months.

£995 or £95 deposit and £39.99 a month for 30 months

Work out
a) how much Sal pays;
b) how much above the cash price she pays.

10 Namita sees the same CD player in two different stores.

£299 inc VAT   £249 ex VAT at 17.5%

Which store offers the better deal and by how much?

11 Emma buys bowls at £20 each and vases at £15 each to sell in her shop.

EMMA'S pots and paintings — £27 — £21

a) Work out the percentage profit on each item.

b) Emma buys 30 bowls and 25 vases and sells them all.

Work out her total profit on the deal.

12 Terry manages a clothing store. He buys shirts at £8 each and trousers at £12 a pair.

a) He prices a shirt at £11.20. Work out his percentage profit on each one.

b) He makes a 50% profit on the trousers. Work out his selling price.

c) Terry buys 50 shirts and 40 pairs of trousers.

Calculate the total cost.

d) He sells 44 shirts and 35 pairs of trousers at the prices in a) and b). He sells the remainder at '25% off' in the sale.

Calculate Terry's total receipts.

e) What is Terry's percentage profit on the whole deal?

Find a shop or an advertisement for computer equipment that quotes prices ex VAT.

Choose 6 items and write down their prices.

Mixed exercise

Sixteen

Probability

Before you start this chapter you should

★ know the meaning of probability

★ know that probability can be represented on a scale from 0 to 1

★ know that a probability of 0 means impossibility

★ know that a probability of 1 means certainty.

Working out probability

Fran is playing ludo. She throws a die and needs a 6 to start. What is the probability that she throws one on her first turn?

Probability is defined as $\dfrac{\text{number of favourable outcomes}}{\text{number of possible outcomes}}$

(This is only true when each outcome is equally likely.)

The probability of throwing a 6 is therefore $\dfrac{1}{6}$

Only 1 favourable outcome, the 6

6 possible outcomes, 1, 2, 3, 4, 5 or 6

How many 6s would you expect Fran to get if she threw the die 600 times?

If Fran throws 1, 2, 3, 4 or 5 she cannot start. The probability of this is $\dfrac{5}{6}$.

5 possible outcomes, 1, 2, 3, 4 or 5

The probability of throwing a 6 can be written as P(6).

That of not throwing a 6 can be written as P(not 6).

You will notice that $\quad$ P(6) + P(not 6) = 1. *Why?*

You can write this as P(not 6) = 1 − P(6).

The probability of something not happening is 1 minus the probability of it happening.

In fact Fran does throw a 6. She passes the die to Pete.

What is the probability that Pete throws a 6?

Pete's throw is not affected by the outcome of Fran's.

Events like this are called **independent**.

16: Probability

1 You select a card at random from a standard pack of 52 cards.

Find the probability that you have selected

a) a heart

b) a red card

c) a court card (king, queen or jack)

d) an ace

e) the queen of hearts.

2 There are 5 coloured balls in a bag. They are yellow, light blue, dark blue, green and red. A ball is taken from the bag at random. Find the probability that

a) it is red

b) it is not red

c) it is a shade of blue

d) it is green or yellow.

3 You throw two dice and add the scores together. Find the probability that

a) the total is 5

b) the total is 12

c) the total is an even number

d) the total is 13

e) you throw a double.

Here is a team game. Each team needs a fair coin, a chart as shown below (extending for a lot more turns!), someone to record on the chart, and a calculator.

The team members take turns to toss the coin. The aim is to throw a head.

The recorder marks the outcome of each toss on a data collection table like this one.

Number of toss	1	2	3	4	5	6	...
Outcome	H	T	T	H	H	H	...
Relative frequency	$\frac{1}{1}$	$\frac{1}{2}$	$\frac{1}{3}$	$\frac{2}{4}$	$\frac{3}{5}$	$\frac{4}{6}$	...

> The relative frequency is the number of Heads so far divided by the number of tosses so far

The results should then be plotted on a chart like this one.

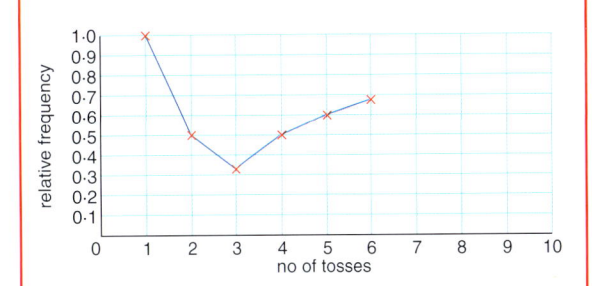

The recorder marks the relative frequency on the chart.

Example: If the 'heads' team scored H, T, T, H, H, H, the recorder would mark the chart
(1, 1) (2, 1/2) (3, 1/3) (4, 2/4) (5, 3/5) (6, 4/6)

The winner is the team with the relative frequency nearest to 0.5.

Would more turns increase your chances of winning?

16: Probability

Estimating probability

In some situations, like throwing dice and tossing coins, you can calculate the probability of any outcome. This is what you did on the last page.

There are other situations where you want to know a probability but cannot work it out from theory. In that case you have to estimate it from data.

Shona spent all of August last year in a small village in Cornwall. On her calendar she put S for sunny or R for rainy against every day.

				5 R	12 S	19 S	26 S
				6 R	13 S	20 R	27 S
				7 S	14 S	21 S	28 S
1 R	8 S	15 S	22 S				29 R
2 S	9 S	16 R	23 S				30 S
3 S	10 S	17 S	24 R				31 S
4 S	11 S	18 S	25 R				

She says

If I choose a day at random the probability of it being wet is $\frac{8}{31}$.

There are 31 days in August

There were 8 rainy days in August

Shona has estimated the probability from the **relative frequency**: the number of times the outcome happened divided by the number of times it could have happened.

? *If Shona collects August rainfall data this year and uses it to do the same calculation, will she get the same result?*

What does Shona mean by 'at random'?

Shona has plans to go back some time in August this year and invite all her friends to an open-air party. Her parents disapprove and tell her that it will almost certainly rain.

She replies 'It is 3 times as likely to be sunny as it is to rain!'

? *Is what Shona says right?*

There are not very many situations where you can work out probability from theory. It is more common to estimate it, but you need more data than Shona collected to have any confidence in your answers.

? *How many years of August weather data would you need to feel confident?*

Do you think you would have got the same results 100 years ago?

People's estimates of probability are not always based on data. Sometimes they are just guesswork, like predicting the result of a sporting fixture.

16: Probability

Exercise

1 Angus is a bird watcher; he keeps a record of all the birds he sees. Over a long period he sees 161 cuckoos, 23 of which are female and the rest are male. Use these figures to estimate:

  a) the probability that a cuckoo chosen at random is female
  b) the probability that a cuckoo chosen at random is male
  c) how many females Angus expects to find among 1000 cuckoos.

2 There is a saying that when a piece of buttered toast falls off the table onto the floor it always lands butter-side-down. Melissa thinks that it is equally likely to land with either side down.

  a) What does Melissa think is the probability of a piece landing butter-side-down?

Melissa hopes to prove her point by carrying out an experiment. She places 20 pieces of buttered toast round the edge of a table and gently tips them off, one at a time. She records the results as u (butter-side-up) and d (butter-side-down). Here are her results.

d d d u d d d u u d d d d d d d u d d u

  b) Use these data to estimate the probability that a piece of toast lands butter-side-down.
  c) What does this suggest about Melissa's original idea?
  d) Does Melissa's experiment prove anything?

3 A doctor is carrying out research into a blood condition. She takes blood samples from 200 people and finds 1 person with the condition. Her assistant uses these results to estimate probabilities.

  a) What is the assistant's estimate of the probability that someone chosen at random has the condition?
  b) What is the assistant's estimate of the probability that someone chosen at random does not have the condition?
  c) How many people would the assistant expect to have the condition in a town with 12 000 inhabitants?

The doctor says that they have not tested enough people to be sure, so she carries out tests on another 1000 people. Of these, 23 have the condition.

  d) Repeat the calculations you did for parts a), b) and c) using all the data now available.
  e) Do your answers show that the doctor was right to ask for more tests?

Estimate the probability that a person selected at random is left-handed. You will have to start by collecting data from a group of people.

16: Probability

Finishing off

Now that you have finished this chapter you should be able to

★ calculate the probability of a particular outcome from theory

★ use data to estimate probability.

★ calculate the probability of an outcome not occurring

Use the questions in the next exercise to check that you understand everything.

Mixed exercise

1 A die is thrown. Find the probability that it comes up

a) an even number

b) a square number

c) a number less than 7

d) a number greater than 7.

e) Draw a number line and mark your answers to parts a), b) and c) on it.

2 A 50p coin and a £1 coin are tossed.

a) Copy and complete this table showing the possible outcomes.

50p	£1
Heads	Heads
Heads	Tails

b) Use your table to write down the probability that the two coins come up

(i) 2 Heads
(ii) 2 Tails
(iii) 1 Head and 1 Tail.

c) Add together your answers to the questions in part b). Why should the total be 1?

d) Are the probabilities the same for any two coins?

e) Another way of showing the possible outcomes from tossing the two coins is this table. Copy and complete it.

	£1 Heads	£1 Tails
50p Heads	HH	HT
50p Tails		

184

16: Probability

Mixed exercise

3 a) Copy and complete this table to show the larger score when 2 dice, one blue and one red are thrown.

b) Work out the probability of each of the possible outcomes, 1, 2, 3, 4, 5 and 6.

c) Show that the probabilities add up to 1.

d) The dice are thrown 360 times. How many times would you expect the outcome to be 1?

Red die score

Blue die score	1	2	3	4	5	6
1			3			
2	2					
3						
4						
5						6
6						

4 Roz is a football enthusiast. She wants to estimate the probability of different numbers of goals being scored by premiership teams. She uses the scores from the matches one weekend.

3–3 1–0 0–0 0–1 1–1
5–0 3–4 5–1 0–1 0–0

a) Use Roz's data to estimate the probability of a team scoring
(i) 0 (ii) 1 (iii) 2 (iv) 3 (v) 4 (vi) 5 (vii) 6 goals

b) Explain why these estimates are obviously not accurate.

c) How could Roz improve her estimates?

d) Are some teams more likely to score many goals than others?

Group activity

One person places a large number of counters of different colours in a bag: for example 20 red, 30 green, 40 blue and 10 yellow, making 100 in all. Everyone knows the total number of counters in the bag but not how many are of each colour.

Each person in the group takes a counter, says what its colour is, and then replaces it in the bag. When this has been done a reasonable number of times the members of the group estimate how many counters of each colour are in the bag.

Answers

Chapter 1: Whole numbers and decimals

Page 4–5: About numbers

1. a) (i) 70 (ii) 50 (iii) 260 (iv) 490
   b) (i) 2100 (ii) 3200 (iii) 4900 (iv) 8000
   c) (i) 2000 (ii) 11 000 (iii) 59 000 (iv) 103 000
2. a) $\frac{1}{8}$ b) $\frac{1}{100}$ c) $\frac{1}{20}$ d) $\frac{1}{40}$
3. a) 4603 b) 20018 c) $3\frac{1}{10}$ (or 3.1)
   d) $2\frac{19}{1000}$ (or 2.019)
4. a) 0.3 b) 0.87 c) 5.7
   d) 2.01 e) 4.113 f) 0.041
5. a) $\frac{1}{5}$ b) $\frac{91}{100}$ c) $4\frac{9}{10}$
   d) $1\frac{87}{100}$ e) $3\frac{1}{4}$ f) $\frac{1}{200}$
6. a) 36.6 °C b) 1.6 m c) 224 ml
7. a) 15.3 b) 11.03 c) £1.35
   d) 18.43 e) 4.236 f) £3.72
8. a) 50 mm b) 0.7 l c) 4000 g d) 1 lb 4 oz
   e) 0.8 cm f) 500 ml g) 1.5 km h) 60 inches
   i) 0.25 kg j) 200 cl k) 40 pints l) 3000 m
9. 7, 14, 21, 28, 35, 42 is one set of six
10. a) 1, 3, 5, 15
    b) 1, 2, 3, 4, 6, 8, 12, 24
    c) 1, 2, 4, 5, 8, 10, 20, 40
    d) 1, 3, 5, 9, 15, 45
    e) 1, 2, 3, 4, 5, 6, 10, 12, 15, 20, 30, 60
    f) 1, 2, 3, 4, 6, 8, 9, 12, 18, 24, 36, 72
    g) 1, 2, 4, 5, 10, 20, 25, 50, 100
    h) 1, 2, 3, 4, 6, 8, 9, 12, 16, 18, 24, 36, 48, 72, 144
11. a) 17 is prime b) 26 is not prime (13 × 2)
    c) 31 is prime d) 39 is not prime (13 × 3)
    e) 73 is prime f) 91 is not prime (7 × 13)
12. a) 25 b) 216 c) 20 d) 900 e) 3
    f) 1024 g) 5 h) 1 000 000 i) 9 j) 3
13. a) 25 k, 7×10^5, 6 million, 30 000 000
    b) 0.009, 5 hundredths, 8×10^{-2}

Page 7: Standard form

1. a) 600 b) 30 000 c) 0.007 d) 0.000 04
   e) 4 500 000 f) 0.0054 g) 9400 h) 0.000 875
   i) 0.016 j) 2 750 000 k) 0.000 083
   l) 10 500 m) 7300 n) 0.000 000 008
   o) 0.4 p) 82 500 000 000
2. a) 4×10^3 b) 8×10^5 c) 3×10^{-3}
   d) 9×10^{-4} e) 2.6×10^4 f) 2.5×10^{-2}
   g) 7.5×10^6 h) 3.7×10^{-5} i) 8.1×10^2
   j) 5.43×10^{-3} k) 9.3×10^{-1} l) 6.4×10^4
   m) 1.6×10^{-2} n) 1.47×10^8 o) 5.07×10^{-1}
   p) 9.04×10^3
3. a) 0.005 b) 4 600 000 000 000
   c) 942 000 000 d) 0.000 0075
4. Pluto (smallest), Mercury, Mars, Venus, Earth, Neptune, Uranus, Saturn, Jupiter (largest)

Page 9: Decimals

1. a) 4.35 b) 4.8 c) 3 d) 3.408
   e) 4.2 f) 12.48 g) 1.04 h) 6.25
2. a) 20 b) 14 c) 44 d) 6.25
   e) 8 f) 3.6 g) $4.\dot{6}$ h) $0.\dot{3}$
3. Malta 81 °F Cyprus 84 °F Tunisia 97 °F
4. a) 48p b) 26p
5. a) 16 litres b) 41 litres c) in a) £11.40; in b) £29.30
6. a) £8000 b) £416 000
7. a) Total is £528.00

Page 11: Imperial and metric units

1. a) 270 l b) 30 or 30.5 cm
   c) 50 miles d) 20 kg e) 6 or 7 gallons
   f) 6 feet 6 inches g) 5 ounces
   h) $\frac{1}{4}$ inch
2. a) 208 km b) 232 km
3. a) 61.8 kg b) 72.7 kg
4. a) 22 lbs b) 8 or 9 pints
   c) 117 or 118 inches d) 7 ounces
5. a) No b) 7 feet 2 inches
   c) 11 gallons d) 22.5 mpg
6. a) 340 g, 11 or 12 ounces
   b) 450 g, 1 lb or 16 oz

Answers

Page 13: Prime factorisation

1. a) 2×7 b) 3×5 c) $2 \times 2 \times 7$
   d) $2 \times 2 \times 3 \times 3$ e) $2 \times 3 \times 5$
   f) $3 \times 3 \times 3$ g) $2 \times 3 \times 3 \times 5$
   h) $2 \times 3 \times 3 \times 7$ i) $2 \times 3 \times 5 \times 5$
   j) $2 \times 3 \times 5 \times 7$ k) $7 \times 7 \times 11$
   l) $2 \times 2 \times 5 \times 7 \times 11$
2. a) 2 b) 3 c) 6 d) 4
   e) 5 f) 1 g) 9 h) 2
   i) 7 j) 11 k) 9 l) 14
   m) 6 n) 5 o) 8 p) 20
3. a) 20 b) 30 c) 8 d) 36
   e) 30 f) 21 g) 54 h) 16
   i) 70 j) 40 k) 60 l) 90
   m) 12 n) 30 o) 72 p) 30
4. a) 35 b) $\frac{14}{35}$ c) $\frac{15}{35}$ d) $\frac{3}{7}$
5. a) 8 b) 5 c) 7

Chapter 2: Shapes and angles

Pages 16–17: Reminder

1. $a = 127°$ $b = 125°$ $c = 76°$
   $d = 123°$ $e = 88°$ $f = 31°$
   $g = 71°$ $h = 109°$ $i = 109°$
2. $p = 30°$ $q = 45°$ $r = 56°$
   $s = 71°$ $t = 60°$ $u = 35°$
   $v = 38°$
3. a) scalene, obtuse-angled
   b) isosceles, right-angled
   c) scalene, acute-angled
   d) isosceles, acute-angled
   e) equilateral, acute-angled
   f) scalene, right-angled
   g) isosceles, obtuse-angled
4. $a = 127°$ $b = 53°$ $c = 127°$ $d = 53°$
   $e = 127°$ $f = 53°$ $g = 62°$ $h = 129°$
   $i = 51°$ $j = 67°$
5. a) r & t, s & q, m & p, o & n
   b) p & s, o & t, m & q, n & r
   c) n & t, p & q

Page 19: Quadrilaterals

1. a) $a = 76°$, $b = 104°$, $c = 53°$, trapezium
   b) $d = 135.5°$, kite
   c) $e = 127°$, $f = 53°$, $g = 53°$, $h = 127°$, parallelogram
   d) $i = 76°$, $j = 104°$, $k = 76°$, rhombus
   e) $l = 117°$, trapezium
   f) $m = 36.5°$, kite
2. a) $a = 60°$, $b = 120°$, $c = 40°$, $d = 20°$, $e = 120°$
   b) $f = 30°$, $g = 50°$, $h = 100°$, $i = 100°$, $j = 30°$
   c) $k = 110°$, $l = 30°$, $m = 40°$, $n = 110°$
   d) $o = 50°$, $p = 50°$, $q = 35°$, $r = s = 95°$, $t = 95°$
3. a) rectangle b) rhombus c) square
4. $x = 115°$, $y = 74°$, $z = 254°$

Chapter 3: Starting algebra

Page 23: Writing things down

1. B, D
2. $2 \times 2 + 2 \times 3$; $2 \times (2 + 3)$
3. a) 11 b) 11 c) 11 d) 2
   e) 77 f) 7 g) 2 h) 24
4. a) 21 b) 16 c) 5 d) 90
   e) 25 f) 27 g) 8 h) 6
5. a) 64 b) 14 c) 400 d) 100
   e) 71 f) 39

Page 25: The language of algebra

1. a) £32 b) £138
2. $8a + 5c + 4s$
3. a) $5a + 6b$, 2 b) $6q - 3p$, 2
   c) $4w + y + 7z$, 3 d) $r - 7s + 1.5t$, 3
   e) $4x + 5y + z$, 3 f) $2k + 12n$, 2
   g) $5b$, 1 h) $12e - 6f$, 2
   i) $3n - 7m + 4p + q$, 4 j) $f + 4d - 19c$, 3
4. a) $13a$ b) $5c$
   c) $9x$ d) $4y$
   e) $4x$ f) d
5. a) $x + 5x + 9x = 15x$ b) $a + 7a - a = 7a$
   $7 - 5 = 2$ $2b + 4b + 6b = 12b$
6. a) $6a$ b) $6a + b$
   c) $7x + 8y$ d) $7x + 4y$
   e) $15n + 6$ f) $13k + 10$
   g) $2y + 11x$ h) $8a + 2c$

187

Answers

Page 27: Substituting into a formula

1. a) (i) 88   (ii) 100   (iii) 106
   b) (i) 48   (ii) 54
2. a) 10   b) 50   c) 64
3. a) 1 m^3 b) $1\frac{9}{16}$ m^3 = 1.56 m^3
4. 80 m
5. 6 m
6. a) 5, moving upwards
   b) 3, moving upwards
   c) 0, at the top
   d) −5, moving downwards

Page 29: Using brackets

1. a) $7x$ b) $13y$
   c) $100n$ d) $30x$
2. a) $3a + 3b$ b) $5c + 5d + 5e$
   c) $10x + 20$ d) $6 + 2y + 2z$
   e) $7f + 21 + 7g$ f) $2x - 10$
   g) $4x - 4y$ h) $2p + 2q - 2r$
   i) $8a + 8d + 8$
3. a) 13   b) 52
   c) $8m + 12$ d) 52
4. a) 8   b) 40
   c) $15x - 10y$ d) 40
5. a) $2x + 5y$, 23   b) $3x + 8$, 20
   c) $12x + 30$, 78   d) $28 + 4x$, 44
   e) $2y + 5x$, 26   f) $8x + 6y$, 50
   g) $6x + 16y$, 72   h) $7x + 8y + 56$, 108
6. a) $3(x + y + z)$ b) $3(x + 2y + 3z)$
   c) $3(2a - b)$ d) $4(2a - b + 4c)$
   e) $6(6p - 2q + 3r)$ f) $25(4u - v - 3w)$

Page 31: Adding and subtracting with negative numbers

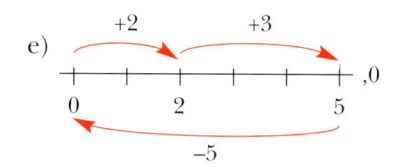

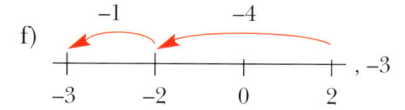

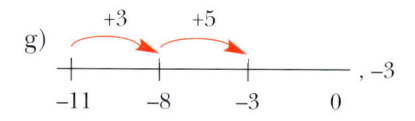

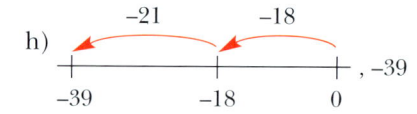

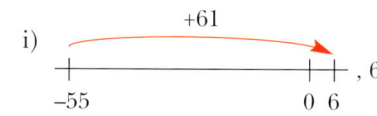

1. a) -4 ; , 2 (number line from 0 to 6)
   b) $+6$; , 2 (number line from −4 to 2)
   c) $+4$; , −2 (number line from −6 to 0)
   d) -6 ; , −2 (number line from −2 to 6)
   e) $+2$, $+3$, -5 ; , 0 (number line from 0 to 5)
   f) -1, -4 ; , −3 (number line from −3 to 2)
   g) $+3$, $+5$; , −3 (number line from −11 to 0)
   h) -21, -18 ; , −39 (number line from −39 to 0)
   i) $+61$; , 6 (number line from −55 to 6)

2. $12 - 1 - 4$, $-1 - 4 + 12$ $12 - 4 - 1$, $-4 + 12 - 1$,
   $-1 + 12 - 4$,

3. a) $2x$ b) $-2x$ c) $-10x$
   d) $4y$ e) $4y$ f) $-4y$
   g) 0   h) y i) $-25y$

4. a) 4   b) −4   c) −20
   d) 12   e) 12   f) −12
   g) 0   h) 3   i) −75

5. a) $5 - 5p$ b) $3q + 1$ c) $p - 6$
   d) $6 - 7p$ e) $3 + 11q$ f) $-p + 3$ or $3 - p$
   g) $4p - 3q$ h) $-18q$ i) 0

6. a) 0   b) 7   c) −5
   d) −1   e) 25   f) 2
   g) −2   h) −36   i) 0

Answers

Chapter 4: Fractions and percentages

Pages 34–35: Reminder

1 a) 3 b) 6 c) 5 d) 14
e) 100 f) 70 g) 13 h) 39

2 a) $\frac{3}{4}$ b) $\frac{1}{4}$ c) $\frac{2}{5}$
d) $\frac{2}{3}$ e) $\frac{3}{5}$ f) $\frac{5}{8}$

3 a) (i) $\frac{1}{5}$ (ii) 5 b) (i) 10 (ii) $\frac{1}{10}$
c) 1 d) You cannot divide by zero.

4 a) 8 b) 9 c) $\frac{3}{4}$ d) $\frac{1}{12}$

5 a) 110 b) 15 c) 60 d) $392\frac{6}{7}$

6 a) $4\frac{1}{2}$ b) $1\frac{5}{8}$ c) $2\frac{2}{5}$
d) $3\frac{2}{3}$ e) $3\frac{3}{4}$ f) $2\frac{1}{6}$

7 a) $\frac{7}{2}$ b) $\frac{35}{8}$ c) $\frac{23}{16}$
d) $\frac{11}{4}$ e) $\frac{16}{3}$ f) $\frac{59}{16}$

8 a) $4\frac{1}{4}$ b) $4\frac{7}{8}$ c) $2\frac{9}{16}$
d) $1\frac{5}{8}$ e) $3\frac{3}{8}$ f) $4\frac{3}{8}$
g) $1\frac{5}{8}$ h) $2\frac{5}{16}$ i) $4\frac{3}{16}$

9 a) $\frac{11}{3}, 4, 4\frac{3}{16}, \frac{17}{4}$ b) $\frac{11}{4}, 2\frac{13}{16}, \frac{23}{8}, 3$

Page 37: Using fractions

1 a) $1\frac{1}{4}$ miles b) $2\frac{3}{4}$ miles
c) 2 miles d) $5\frac{3}{4}$ miles

2 a) 40 b) $3\frac{3}{4}$ pies
3 a) 14 b) 1545
4 a) £23.33 b) No
5 a) 48 b) 32 c) 160 d) 120
6 60

Page 39: From fractions to percentages

1 a) 32%, 0.32 b) 55%, 0.55
c) 12.5%, 0.125 d) 67.5%, 0.675
e) 71.6% 0.716 f) $46.\dot{6}$%, $0.4\dot{6}$

2 a) Knight $\frac{4}{5}$; Chequer $\frac{12}{25}$; Domino $\frac{7}{10}$;
b) Knight 80%; Chequer 48%; Domino 70%

3 a) 47.5% b) 31.25% c) 21.25%
4 a) 125 b) 56% c) 52%

Page 41: Using percentages

1 a) 240 b) 150 c) 315.5
d) 87 e) 25 f) 27.9
2 a) £1500 b) £187.50
3 a) £8240 b) £8569.60
4 6089
5 £396
6 a) 12.5% b) 7.5%
7 a) May + 15%; June –4%;
July + 5%; August + 3.5%
b) 4.3%

Page 43: Making comparisons

1 a) 0.8 b) 0.44 c) 0.375
d) $0.\dot{6}$ e) $0.1\dot{6}$ f) $0.58\dot{3}$
2 0.09, 0.83, $\frac{5}{6}, \frac{21}{25}, \frac{17}{20}$
3 A
4 a) Northhill 79.5%; Heartland 54.9%;
Southdown 78.8%
b) The Heartland centre is less successful than the other two.
5 a) A 4.5% B 4.9% C 4.6%
b) A
c) Breakdown or older machine

Chapter 5: Area and volume

Page 47: Parallelograms and trapezia

1 a) 24 cm², 20 cm b) 38 cm², 30 cm
c) 40 cm², 35 cm
d) 20 cm² e) 112.5 cm² f) 126 cm²
2 a) 42 cm² b) 92 cm² c) 63 cm² d) 60 cm²
3 10 000 cm² (or 1 m²)

Answers

Page 49: Circumference and area of a circle

1 a) 5π cm, 15.7 cm b) 8π cm, 25.1 cm
 c) 6π cm, 18.8 cm

2 a) 4π cm^2, 12.6 cm^2 b) 49π cm^2, 153.9 cm^2
 c) $\frac{121}{4}\pi$ cm^2, 95.0 cm^2

3 a) $\frac{25}{\pi}$ cm, 7.96 cm b) $\frac{12}{\pi}$ cm, 3.82 cm
 c) $\frac{34}{\pi}$ cm, 10.82 cm

4 a) $\sqrt{\frac{30}{\pi}}$ cm, 3.09 cm b) $\sqrt{\frac{56}{\pi}}$ cm, 4.22 cm
 c) $\sqrt{\frac{112}{\pi}}$ cm, 5.97 cm

5 8π cm, 25.1 cm

6 a) $\frac{12}{\pi}$ m, 3.82 m b) $\frac{144}{\pi}$ m^2, 45.84 m^2

7 a) $(12 + 2\pi)$ cm^2, 18.3 cm^2
 b) $(26 + 4\pi)$ cm^2, 38.6 cm^2
 c) $(28 + 4\pi + \frac{49\pi}{4})$ cm^2, 79.1 cm^2
 d) $(100 - 25\pi)$ cm^2, 21.5 cm^2

8 a) $\frac{25\pi}{4}$ cm^2, 19.6 cm^2 b) $\frac{25}{2}$ cm^2, 12.5 cm^2
 c) $\frac{25\pi}{4} - \frac{25}{2}$ cm^2, 7.1 cm^2

Page 51: Volume of a prism

1 a) 45 cm^3 b) 100 cm^3 d) 18 cm^2 e) 62.8 cm^3
 d) 30 cm^3 e) 35.3 cm^3

2 870 m^3

3 795

Page 53: Surface area of a prism

1 a) 114 cm^2 b) 232 cm^2 c) 125.7 cm^2
 d) 108 cm^2 e) 99.8 cm^2

2 556 m^2

3 928.3 m^2

Page 55: More about prisms

1 $a = 1.99$ cm $b = 4$ cm $c = 2$ cm
 $d = 2.44$ cm $e = 15.92$ cm $f = 6$ cm

2 a) 150.5 cm^2 b) 146 cm^2 c) 108 cm^2
 d) 98.7 cm^2 e) 84.9 cm^2

3 a) 8.84 cm b) 3.26 cm

4 2.78 m

Page 57: Using dimensions

1 (i) a) 3 dimensions; x, x, y
   b) volume
   c) C
 (ii) a) 1 dimension; $x + y$
   b) length
   c) A
 (iii) a) 2 dimensions; y, y
   b) area
   c) E
 (iv) a) 2 dimensions; x, y
   b) area
   c) A
 (v) a) 2 dimensions; $y, x + y$
   b) area
   c) D
 (vi) a) 3 dimensions; x, x, y
   b) volume
   c) F
 (vii) a) 2 dimensions; x, y
   b) area
   c) B
 (viii) a) 1 dimension; y
   b) length
   c) E

2 a) volume
 b) area
 c) length
 d) area
 e) not a real formula
 f) area
 g) not a real formula
 h) volume

Answers

Chapter 6: Using symbols

Page 61: Being brief

1. a) 5^3 b) 8^5 c) 10^3 d) 7^9
   e) 10^8 f) 10^1 g) x^4 h) $4y^2$
   i) $9n^3$ j) $3a^5$
2. a) $2 \times 2 \times 2$ b) $6 \times 6 \times 6 \times 6$
   c) $d \times d \times d$ d) $n \times n \times n \times n \times n$
   e) $5 \times c \times c$ f) $7 \times g \times g \times g \times g$
   g) $10 \times z$ h) $2 \times x \times x \times x \times x \times x \times x$
   i) $2 \times m \times m \times m \times m \times m$
   j) $8 \times u \times u \times u$
3. a) 3^4 b) 6^{10} c) 4^8 d) 10^{17}
   e) a^5 f) v^{11} g) s^5 h) d^8
   i) x^7 j) n^8
4. a) $6x$ b) $20x^2$ c) $6a^3$ d) $6g^2$
   e) $6x^2$ f) $20x^2$ g) $3p^3$ h) $7y^5$
   i) $2n^4$ j) $90m^6$
5. 2^7 is the biggest.
   $2 \times 7, 7^2, 2^7$
6. 4^7 is the biggest.
   $1^{10}, 10^1, 9^2, 2^9 = 8^3, 7^4, 3^8, 6^5, 5^6, 4^7$
7. a) 3^2 b) 2^6 c) 5^5 d) x^0
8. a) 8 b) 104 c) 4
9. a) 16 b) 2000 c) 0

Page 63: Using negative numbers

1. a) -5 b) 5 c) -5
   d) 5 e) -7 f) 3
   g) $-12r$ h) $-14s$ i) $99t$
2. a) 5 b) 17 c) -21 d) 13
   e) -5 f) 8 g) 24 h) -18
   i) 999 j) 0
3. a) $9 + m$ b) $9 - m$ c) $3 - a$
   d) $3 + 2a$ e) $4 - 2n$ f) $7 + 2n$
4. a) 4 b) -8 c) 8 d) -4
5. 2035
6. a) 7 b) 1 c) 10 d) -2

Page 65: Simplifying expressions with negative numbers

1. a) -8 b) -8 c) 8 d) 24
   e) -24 f) 20 g) 9 h) 64
   i) -6 j) 6 k) -6 l) -11
   m) $-10x$ n) $8x$ o) $3x$
2. a) 9 b) 11 c) -1
   d) 0 e) 1 f) 19
3. a) $8z + 28$ b) $8z - 28$
   c) $-8z - 28$ d) $-8z + 28$
   e) $60 + 20a$ f) $-60 - 20a$
   g) $-60 + 20a$ h) $-7e - 14$
   i) $-4 + 8y$ j) $-4 - 8y$
4. a) 16 b) 8 c) -4 d) 3
   e) 17 f) 11 g) -3 h) 6
5. a) $27 + 2x$ b) $15 - 2x$ c) $10n - 10$
   d) $2n + 10$ e) $13d + 4$ f) $a + 30b$
   g) $15 - 10t$ h) $4 + 6c$ i) 18
6. a) -10 b) 12 c) -70

Chapter 7: Data handling

Page 69: Pie charts

1. Rent 60°, Gas/Electricity 45°, Food 110°, Household 60°, Personal 30°, Clothing 20°, Fares 15° Other 20°.
2. Finance 144°, Sales 90°, Management 72°, Administration 27°, Market Research 18°, Marketing 9°.
3. a) 180 altogether
   b) France 50; Switzerland 30; Austria 10; Germany 50 and Poland 40.

Page 71: Stem-and-leaf diagrams

1. b) 30 c) 73 d) 3
2. b) 21 c) 69 d) 9
3. b) 6 c) 37 d) 3 e) 14
4. b) 180 mm, 257 mm c) 150 mm, 245 mm
   d) Jane's flowers are taller: fertiliser is effective.

Page 73: Moving averages

1. a) 3
   b) 5, 4.67, 4.33, 4.67, 3.33, 4.33, 3.67, 6, 5.67, 8, 5.33, 5, 3, 5.33, 4.33, 4.67
2. a) 12 b) 7 c) 6
3. a) 77 b) £100 c) 79, £84
4. b) 3.15, 3.25, 3.35, 3.45, 3.55, 3.65, 3.75, 3.85, 3.95
   d) Rising e) 2600, 4400
5. b) 31.33, 33, 36.33, 35.67, 36.3, 46.33, 47.67, 48.67, 35.33, 36
   d) 3 e) Illness

Answers

Page 75: Bivariate data

1 a) Ask your teacher to check your scatter diagram.
There is a strong positive correlation between weight and length.
There is no obvious difference between girls and boys.

2 Ask your teacher to check your scatter diagram.
There is a strong positive correlation up to 100 kg per ha, then a strong negative correlation if fertiliser is increased.

Page 77: Line of best-fit

1 a) Ask your teacher to check your scatter diagram.
b) 19.5 N
c) 5.3 kg

2 a) Ask your teacher to check your scatter diagram.
b) Strong positive correlation
c) Ed is wrong. Should be 67.5
e) 0.133 g

3 a) Ask your teacher to check your scatter diagram.
b) 7.78 hours, 58 visitors
c) 57

Chapter 8: Graphs

Page 81: Looking at graphs

1 a) b)

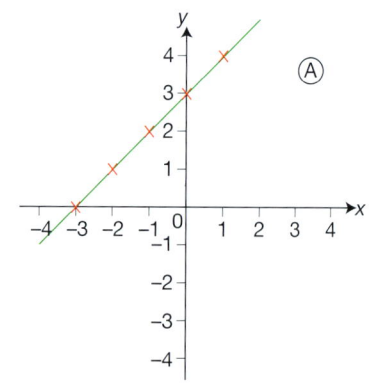

c) Add 3 to x
d) $y = x + 3$

a) b)

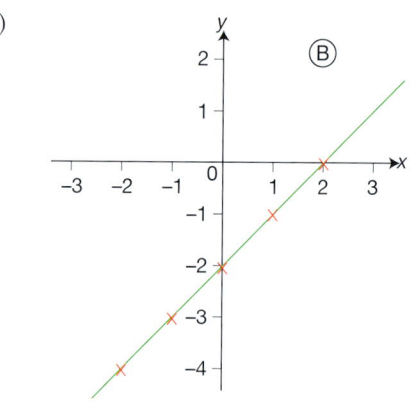

c) Multiply x by 3
d) $y = 3x$

a), b)

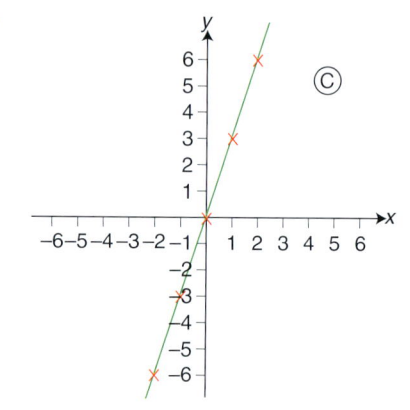

c) Subtract 2 from x
d) $y = x - 2$

2 a)

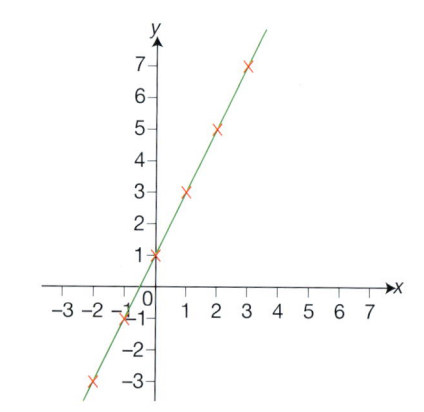

b) It increases by 2 each time.

3 d) $H = 150 + 20D$
a)

D	0	5	10	15
H	150	250	350	450

Answers

b)

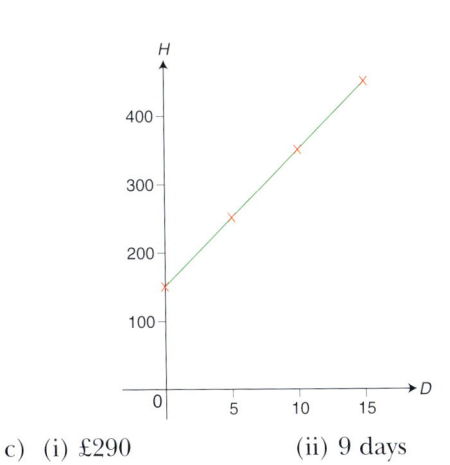

[Graph showing H vs D, line passing through points from (0,150) rising to approximately (15,450)]

c) (i) £290   (ii) 9 days

Page 83: Gradients and intercepts

1 a)

x	−2	−1	0	1	2	3	4
y	−10	−7	−4	−1	2	5	8

b), c) Ask your teacher to check your scales and graph.
d) Gradient = 3   e) Intercept = −4

2 a)

x	−2	−1	0	1	2	3	4
y	$\frac{1}{2}$	$\frac{3}{4}$	1	$1\frac{1}{4}$	$1\frac{1}{2}$	$1\frac{3}{4}$	2

b), c) Ask your teacher to check your scales and graph.
d) Gradient = $\frac{1}{4}$ e) Intercept = 1

3 a) 1, 5, $y = x + 5$ b) 2, −2, $y = 2x − 2$
c) $-\frac{2}{3}$, 2, $y = -\frac{2}{3}x + 2$ d) $-\frac{1}{2}$, −1, $y = -\frac{1}{2}x − 1$

Page 85: Obtaining information

1 a) 25p   b) £2.50
c) A (2, 5) B (6, 10) C (6, 5)
d) AC = 4 miles, BC = £5   e) £1.25 per mile
2 a) £25   b) £350
c) A (1, 500) B (6, 1250); AC = 5 thousand, BC = £750
d) £150   e) £1850
3 a) £100   b) £35 per day
c) $C = 100 + 35x$ d) £800

Page 87: Travel graphs

1 a) 12.30 p.m., $1\frac{1}{2}$ hours b) 15 minutes
c) 2 hours, she stopped for 1 hour.
d) the same speed
2 a) 3.30 p.m., 20 miles   b) 15 miles per hour
c) He had a break for lunch then went on more slowly.
d) −10 miles per hour. He was going back home.
e) They meet 16 miles from home, at 3.54 p.m.
3 a) Ask your teacher to check your graph.
b) 50 miles per hour, 48 miles per hour, 70 miles per hour
c) At 12.55, about 42 miles from Oxford.

Page 89: Curved graphs

1 a)

t	0	1	2	3	4
$20t$	0	20	40	60	80
$-5t^2$	0	−5	−20	−45	−80
h	0	15	20	15	0

b) Ask your teacher to check your graph.
c) 0.5 s and 3.5 s. The ball is 10 m above the ground both on the way up and on the way down.
d) If the graph were extended, the ball would seem to continue falling below the ground.

2 a)

x	0	1	2	3	4	5
−4	−4	−4	−4	−4	−4	−4
$+5x$	0	5	10	15	20	25
$-x^2$	0	−1	−4	−9	−16	−25
y	−4	0	2	2	0	−4

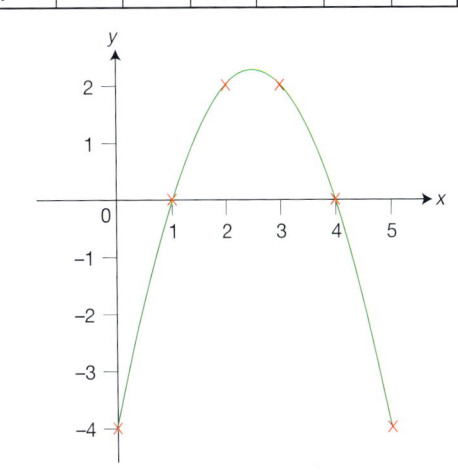

b) [Parabola graph opening downward with vertex near (2.5, 2.25), crossing x-axis near x=1 and x=4]

c) 2.25
d) 1.4, 3.6

Answers

3 a)

x	−2	−1	0	1	2
x^3	−8	−1	0	1	8
$-2x$	+4	2	0	−2	−4
y	−4	1	0	−1	4

b) [graph]

c) 1.4, 0, −1.4
d) Rotational symmetry about O.

Chapter 9: Ratio and proportion

Pages 92–93: Revision exercise

1 a) 10   b) 3   c) 4   d) 20
   e) 18   f) 16   g) 2   h) 50
   i) 3   j) 3,5
2 a) 6   b) 4   c) 2
3 a) 3 : 4   b) 4 : 1   c) 1 : 3   d) 5 : 1
   e) 5 : 3   f) 5 : 8   g) 2 : 5   h) 3 : 2
   i) 4 : 3   j) 3 : 7   k) 5 : 2   l) 2 : 3 : 5
4 a) 4   b) 120   c) 4 : 1
5 a) 4 : 1   b) 1 : 20   c) 7 : 10
   d) 5 : 1   e) 2 : 1   f) 1 : 50 000
   g) 3 : 2   h) 1 : 4   i) 5 : 3
6 a) 3 : 4   b) 240
   c) 180   d) 240 : 180 (or 4 : 3)
   e) the last answer is the reverse of answer a).
7 a) 60 ml   b) 20 ml : 1 l (or 1 : 50)   c) 1 : 50
8 a) $\frac{2}{5}$ b) 60%   c) £15 000
9 a) £180, £120   b) £140, £70
   c) £3750, £2250   d) £187.50, £562.50
   e) £260, £325   f) £2.50, £1

Page 95: Using ratio

1 a) 6 : 5   b) $\frac{6}{11}$ c) 48
   d) $\frac{5}{11}$ e) 40
2 a) 3 : 5   b) Charlotte £150, Thomas £250
   c) £640   d) Charlotte £234, Thomas £390
3 a) 2.5 tonnes   b) 2 : 3
4 a) Home 33 000, away 5500
   b) 3600   c) 13 : 1

Page 97: Unitary method

1 a) 6 pack   b) 6 pack   c) 500 g pack
2 a) 96.5 g   b) 579 g
3 a) £4.80   b) £182.40   c) 42
4 a) 255 miles   b) 14   c) No
5 a) £280   b) 140   c) 4   d) 80
6 a) 100 g flour, 40 g sugar, 40 g butter
   b) 500 g   c) 10

Page 99: Changing money

1 a) 7400   b) 17 575   c) £100   d) £43.24
   e) £4.05
2 a) 1785   b) £306
   c) Jacket £31.09   Camera £109.24
   d) £20.84
3 a)

Rand	£	Rand	£
1	£0.06	15	£0.94
2	£0.13	20	£1.26
5	£0.31	50	£3.14
10	£0.63	100	£6.29

   b) (i) 7 = 5 + 2
      (ii) 45 = 50 − 5
   c) (i) 7 rand = £0.44
      (ii) 45 rand = £2.83

Page 101: Distance, speed and time

1 a) 100 km   b) 105 km   c) 1980 km   d) 20 km
2 a) 56 km/h   b) 64.8 km/h
   c) 84 km/h
3 a) $2\frac{1}{2}$ h   b) $3\frac{3}{4}$ h   c) 1h 40 min
   d) 2 h 6 m 36 s
4 a) 1240   b) 80 km/h
   c) 64 km/h   d) 35 minutes

Answers

Chapter 10: Grouped data

Page 105: Grouping continuous data

1 a)

Mileage (thousands)	0≤m<10	10≤m<20	20≤m<30	30≤m<40	40≤m<50
Frequency	1	2	5	2	2

b) Ask your teacher to check your histogram.
c) (i) Total price: A £118,000, B £123,000.
   B is best.
   (ii) Sell 2 highest mileage to A and rest to B. £125,000

2 a)

Reading	70≤m<71	71≤m<72	72≤m<73	73≤m<74	74≤m<75
Frequency	0	1	4	5	10

Reading	75≤m<76	76≤m<77	77≤m<78
Frequency	4	4	2

b) Ask your teacher to check your histogram.
c) less than 72 or more than 78, i.e. only 1 reading

Page 107: Grouping rounded data

1 a) 5.5 and 10.5 b) 8
c) just under 20.5 minutes d) 0.5 mins
e) Ask your teacher to check your histogram.

2 a) 70.5 and 140.5 b) 105.5
c) just under 280.5 g
d) Ask your teacher to check your histogram.

3 a)

Age in years	16–20	21–25	26–30	31–35	36–40	41–45	51–55	66–70
Frequency	2	11	6	5	2	2	1	1

b) Ask your teacher to check your histogram.
c) add 1 to each of the labels on the x axis

4 a) 15 sq units (each sq unit represents 1 bus)
b) 15 sq units. The area of the histogram is the same as the area under the frequency polygon.

Page 109: Frequency polygons

1 a),b) Ask your teacher to check your frequency polygons.
c) Owens Academy had more completed laps and a higher mode than Avonford High.

2 a) There are more females than males on the trip. The females tend to be younger.
b) Ask your teacher to check your frequency polygons.

Page 111: Mean, median and mode of grouped data

1 a) 6 – 10 kg b) 8 kg c) 430 kg d) 8.6 kg

2 a) £51 000 – £60 000 b) £105 500
c) $\frac{£7036500}{143} = £49206$

3 a) 90 – 94 mins b) 92.25 mins
c) 92.5 mins (approx 25 mins added to total)
d) modal class still 90 – 94 mins

Page 113: Cumulative frequency

1 a) Ask your teacher to check your cumulative frequency curve.
b) 11 years
c) 510

2 a) Ask your teacher to check your cumulative frequency curve.
b) Men 38.3, women 35
c) Women in the competition tend to be younger than men in the competition.

3 a) Ask your teacher to check your graph.
b) The graph is symmetric. The number dead gives a 'less than' cumulative frequency curve and the number alive gives a 'more than' cumulative frequency curve. They intersect at the median value.

Page 115: Quartiles

1 a) Ask your teacher to check your cumulative frequency curve.
b) Median 371, inter-quartile range 373 – 367 = 6
c) 6

2 a) Ask your teacher to check your cumulative frequency curve.
b) 14
c) 21 – 6.5 = 14.5
d) This is a young person's disease.

Page 117: Box-and-whisker diagrams

1 Median 9, mode 9, quartiles 5 and 17
2 b) LQ 25, median 32, UQ 37
3 c) LQ 1.75, median 2.55, UQ 3.575
4 b) Trained: LQ 125, median 129, UQ 132
   Untrained: LQ 139, median 144, UQ 149
d) Trained employees can do the job in less time but their range of times is wider.

Answers

Chapter 11: Equations

Page 121: Using equations

1 a) $n = 54$ b) $v = 15$ c) $c = 8$
  d) $f = 42$ e) $x = 10$ f) $y = 10$

2 a) $x = 3$ b) $y = 6$
  c) $z = 5$ d) $x = 5$

3 a) $y = 10$ b) $x = 2.4$ c) $x = 2$ d) $c = 4.1$
  e) $y = 10$ f) $n = 0.9$ g) $x = 1$ h) $n = 78$
  i) $m = 11$ j) $u = 3$ k) $t = 10$ l) $x = 2$
  m) $x = 2.5$ n) $x = 1.3$ o) $y = 0$

4 a) $x + 120 = 180$ b) $2y + 40 = 180$
    $x = 60$ $y = 70$
  c) $5z = 180$
    $z = 36$

Page 123: Solving equations

1 a) $h = 8$ b) $n = 3$ c) $x = 17$
  d) $y = \frac{1}{2}$ e) $k = 1$ f) $x = 6\frac{1}{2}$
  g) $x = 10\frac{1}{2}$ h) $y = 1.2$ i) $x = -3$

2 a) $x = 1$ b) $a = -2$ c) $x = 2$ d) $y = 1$
  e) $d = 6\frac{2}{3}$ f) $x = 4$ g) $k = 5$ h) $k = -5$

3 a) $a = 6$ b) $t = 7$ c) $y = 4$
  d) $c = \frac{1}{2}$ e) $x = 0$ f) $d = -1$

4 a) $c = 3.75$ b) $y = 1.286$
  c) $x = 0.267$ d) $p = 3.273$
  e) $x = 1.857$ f) $y = 10$
  g) $b = 1.136$ h) $a = 0.3$

5 a) 20 b) 86 c) $F - 32 = \frac{9C}{5}$
          $F = \frac{9C}{5} + 32$

Page 125: Equations with fractions

1 a) $x = 10$ b) $x = 24$ c) $x = 10\,000$
  d) $x = -4$ e) $x = 4$ f) $x = 8$

2 a) $\frac{x}{12} = 5 - 2$ b) $x = 36$

3 a) $x = 24$ b) $x = 3$ c) $x = 12$ d) $x = 11$
  e) $x = 12$ f) $x = 9$

4 a) $x = 11$ b) $x = -2$ c) $x = 1$ d) $x = -9$
  e) $x = -2$ f) $x = \frac{1}{2}$ g) $x = 1$
  h) $x = 2$ i) $x = 4$

5 a) $x = 15$ b) $x = 5$ c) $x = \frac{1}{2}$ d) -9

6 a) (i) 40 (ii) $R = \frac{240}{6}$
  b) $R = \frac{V}{I}$

Page 127: Changing the subject of a formula

1 a) $x = y - 4$ b) $x = y - 20$ c) $x = y - a$
  d) $x = y - 3$ e) $x = y - 13$ f) $x = y - c$
  g) $x = y + 5$ h) $x = y + 11$ i) $x = y + b$
  j) $x = 6 - y$ k) $x = 1 - y$ l) $x = d - y$

2 a) $x = \frac{1}{2}y$ b) $x = 10y$ c) $x = \frac{y}{a}$
  d) $x = 4y$ e) $x = 10y$ f) $x = by$
  g) $x = \frac{4}{3}y$ h) $x = \frac{3}{5}y$ i) $x = \frac{b}{a}y$
  j) $x = \frac{5}{4}y$ k) $x = \frac{2}{11}y$ l) $x = \frac{b}{a}y$
  m) $x = \sqrt{p}$ n) $x = \sqrt{\frac{q}{2}}$ o) $x = \sqrt{(l + m)}$
  p) $x = s^2$

3 a) $t = \frac{x + 3}{2}$ b) $t = \frac{y - 4}{3}$ c) $t = \frac{p - 6}{2}$
  d) $t = 4 - c$ e) $t = \frac{6 - z}{2}$ f) $t = \frac{s - a}{2}$
  g) $t = \frac{x + c}{5}$ h) $t = \frac{n + 3x}{7}$ i) $t = \sqrt[3]{p}$

4 a) $u = v - at$ b) $l = \frac{p - 2b}{2}$ c) $x = \frac{V + 9y}{4}$
  d) $t = \frac{v - u}{a}$ e) $x = \sqrt{(a - b)}$ f) $x = (a - b)^2$

5 a) $x = \frac{p - 2y}{2}$ or $\frac{p}{2} - y$
  b) $x = \frac{V - 12r}{12}$ or $\frac{V}{12} - r$
  c) $x = \frac{8 - s}{4}$ d) $x = \frac{4a - y}{4}$ or $a - \frac{y}{4}$

6 a) $l = \frac{A}{b}$ b) $h = \frac{V}{lb}$ c) $R = \frac{V}{I}$
  d) $d = \frac{c}{\pi}$ e) $r = \frac{c}{2\pi}$ f) $P = \frac{100I}{r}$
  g) $T = \frac{100I}{PR}$ h) $R = \frac{100I}{PT}$ i) $d = \sqrt{\frac{4A}{\pi}}$

Answers

Chapter 12: Approximations

Page 131: Decimal places

1. a) 3.1 b) 3.14 c) 3.142
2. a) 4.5 b) 4.47 c) 4.472
3. a) 1 b) 2
4. a) Approximately 5.38 to the nearest hundreth; 5.4 to the nearest tenth
   b) Approximately 8.72 to the nearest hundreth; 8.7 to the nearest tenth
5. a) 0.167 b) 0.083 c) 0.667 d) 0.429
6. 5.52
7. a) 5.8 cm, 2.3 cm b) 13.3 cm^2
8. 7.86
9. a) 45.8 m b) 167.3 m^2

Page 133: Significant figures

1. a) 77 000 b) 42.2 c) 5400 d) 758 000
   e) 4000 f) 62.0 g) 6.7 h) 50 000
   i) 33.8 j) 0.005
2. a) Approximately 37.73 to 4 significant figures; 37.7 to 3 significant figures
   b) Approximately 0.2156 to 4 significant figures; 0.216 to 3 significant figures
   c) Approximately 1.444 to 4 significant figures 1.44 to 3 significant figures
3. a) 38 b) 0.22 c) 1.4
4. 26 000
5. a) 370 m b) 8300 m^2
6. a) £90 000 b) £360 000
7. a)

Day	Thursday	Friday	Saturday
Tickets sold	257	319	348
Income	£1090	£1360	£1480

   b) £3900

Page 135: Estimating costs

1. a) £40 b) £36 c) £35
   d) £25 e) £165 f) £160
2. a) 200 b) £2000
3. £80
4. a) £80 or £84 b) £60
   c) £240 d) £600

Page 137: Using your calculator

1. a) 280 m^2 b) 400 cm^2 c) 48 cm^2
2. a) £2000 b) £100 000
3. a) 445.60 ringgits b) 4971.20 baht
   c) 116.80 Fiji dollars d) 208 Australian dollars
4. a) 14 000
   b) 160 000, 520 000, 200 000, 240 000, 280 000
   c) not a typical attendance, e.g. high for a local derby
5. a) 24 m b) 12 m c) 25 m
6. a) 48 m^2 b) 12 m^2 c) 75 m^2

Page 139: Using fractions and percentages

1. a) 200 b) 25 c) 180 d) 200
   e) 250 f) 600 g) 300 h) 360
2. a) 10 000 b) 7500 c) 5000
3. a) £6000 b) £3000 c) £2000 d) £1000
4. a) 300 b) 25 c) 200 d) 40
   e) 180 f) 250 g) 150 h) 60
5. a) £800 b) £270 c) £700
6. a) £12 b) £26 c) £45

Page 141: Errors

1. 3950 ft   3850 ft
2. 350 miles   280 miles
3. a) 18.5 cm   17.5 cm b) 23.5 cm   22.5 cm
   c) 606.5 cm   577.5 cm
4. a) 247.5 miles b) 157.5 miles
5. Smallest 344.375 m^3, largest 569.625 m^3

Answers

Chapter 13: Practical statistics

Pages 144–145: Revision exercise

1. a) numerical, continuous
   b) numerical, continuous
   c) categorical
   d) categorical
   e) numerical, discrete
2. a) 2.30 : 2 miles, 14 horses
   3.00 : 1 mile 1 furlong, 6 horses
   3.30 : 6 furlongs, 12 horses
   4.00 : 2 miles 7 furlongs, 8 horses
   4.30 : 2 miles, 15 horses
   5.00 : 7 furlongs, 12 horses
   b) 4.30   c) 4.00
3. a)

Cause	Area (hectares)
Slag heaps	13 500
Excavations	9000
Disused railway lines	10 350
Military	4 500
Other	7 650

   b) Ask your teacher to check your bar chart.
   c) 124 200 sheep could be reared.
4. a) Ask your teacher to check your drawings.
   b) 21 – 25   c) Yes, it's possible
5. a) Ask your teacher to check your bar chart.
   b) 53.333%

Pages 146–147: Reminder

1. a) 9, 9, 8, 11   b) 5.4, 5, 4, 4
   c) 2.8, 2.5, 2, 5   d) 14, 10, 10, 23
2. a) 1   b) 10   c) 9
3. £672
4. a) 1.6 m b) 1.7 m c) 1.71 m d) 1.6 m
   e) The mode is still 1.6; the median is still 1.7.
5. a)

Number of cars per week	0	1	2	3	4
Frequency	1	4	7	5	3

   b) 4
   c) 2
   d) Ask your teacher to check your chart.
   e) It is the tallest column.

Page 149: Which average?

1. a) Ask your teacher to check your chart.
   b) Mean 2.3. mode 1, median 2, range 9
   c) Mean 1.95, mode 1, median 2, range 5
2. a)

Number of occupants	1	2	3	4
Frequency	20	10	5	15

   b) Most cars had only a driver, next most common had 4
   c) Mean 2.3, mode 1, median 2
   d) Mode and median
   f) 1.3
3. a) 3   b) 2.5
   c)

Number of cars sold per week	1	2	3	4
Frequency	15	15	25	5

   d) 2.333
   e) Wants a high total, so a high mean.

Page 151: Making comparisons

1. a) Shop A, 6–10. Shop B, 16–20
   b) Ask your teacher to check your frequency polygons.
   c) Fewer items are bought in shop A than in shop B (lower modal class) but the range is the same.
   d) A is probably a smaller shop because most people just buy a few items. B is probably a supermarket.
2. a) Ask your teacher to check your cumulative frequency chart.
   b) Northbound: median 3.25 mins, IQ range
      4 mins 40 secs – 2 mins = 2 mins 40 secs
      Southbound: median 7 mins, IQ range
      9 mins – 5 mins = 4 mins
   c) Southbound traffic has longer delays which could be helped by setting traffic lights with longer at green for Southbound traffic.
   d) Yes. If this is morning rush hour then the traffic pattern would be reversed in the evening rush hour.

Answers

Chapter 14: Finding lengths

Page 157: Scale drawings

1. a) 20 cm by 8 cm
   b) Length 4 cm, height 2.5 cm
2. b) 19.6 cm, 3.92 m in reality
   c) About 4.2 cm, 84 cm in reality
3. a) 0.0075 mm
   b) 40 cm

Page 159: Using bearings

1. a) A 030°, B 135°, C 315°, D 260°
   b) A 15 km, B 20 km, C 30 km, D 25 km
2. a) 225°   b) 45°   c) 280°
   d) 100°   e) 20°   f) 200°
   g) There is a difference of 180° between each pair of bearings.
3. 1.8 or 1.9 km and 5.0 or 5.1 km

Page 161: Pythagoras' rule

1. a) 8.06 cm   b) 8.49 cm   c) 13.60 cm
   d) 13.45 cm   e) 8.16 cm   f) 9.17 cm
2. 192 m
3. 28.6 km
4. a) 3 units   b) 2 units   c) 3.61 units
5. a) 5.39 units b) 4.24 units c) 5.10 units
   d) 4.47 units

Page 163: Finding one of the shorter sides

1. a) $\sqrt{39}$ cm   b) $\sqrt{80}$ cm   c) $\sqrt{20}$ cm
   d) 8.02 cm   e) 4.08 cm   f) $\sqrt{50}$ cm
2. 2.44 m
3. a) $\sqrt{32}$ cm   b) $2\sqrt{32}$ cm^2
4. a) 3 cm   b) 6 cm^2 c) 2.4 cm
   d) AN = 3.2 cm, CN = 1.8 cm

Chapter 15: Money

Page 167: Simple interest

1. a) £30   b) £330   c) £235.60
   d) £975   e) £17   f) £6850
2. a) Grant, Southern Building Society, £94.50
      Maxine, Eastern Building Society, £1072.50
      Scott, Western Building Society, £187.50
   b) £18.75
3. a) £168.75   b) £475   c) £84.38   d) 4.5%
   e) Mohan £33.75, Alka £16.87

Page 169: Compound interest

1. a) £66.20   b) £102.50   c) £18.54
   d) £4277.92   e) £8082.16   f) £964.88
2. a) £1770   b) £1786.52
3.

Year	Amount at start of year	Interest
1	£800	£800 × 8/100 = £64
2	£864	£864 × 8/100 = £69.12
3	£933.12	£933.12 × 8/100 = £74.65
4	£1007.77	£1007.77 × 8/100 = £80.62
5	£1088.39	£1088.39 × 8/100 = £87.07

Year	Amount at end of year
1	£800 + £64 = £864
2	£864 + £69.12 = £933.12
3	£933.12 + £74.65 = £1007.77
4	£1007.77 + £80.62 = £1088.39
5	£1088.39 + £87.07 = £1175.46

4. a) £1368   b) £1361.07   c) 4 years

Page 171: Bills

1. a) 605   b) £79.86   c) £94.66
2. a) 82   b) £44.69   c) £58.89
3. a) 664   b) £50.46   c) £62.96   d) 500
4. a) £96.50
   b) Yes. The bill is £92.98 a saving of £3.52
   c) A person using fewer than 4200 units per quarter

Page 173: Hidden extras

1. a) £927, £178   b) £1426.64, £231.64
2. a) £950.40   b) £1440
   c) £330   d) £546.75
3. a) £152.75   b) £434.75
4. a) A by £9.17   b) B by £22.33

Page 175: Profit and loss

1. a) Profit 52%   b) Profit 37.5%
   c) Loss 30%   d) Loss 14.4%
2. a) £72   b) £86.80   c) £68   d) £14.25
3. a) £180   b) 30%
4. a) Jackets £6, Trousers £4
   b) Jackets 12%, Trousers 20%
5. a) £330   b) 49.3%
6. a) £115   b) £25   c) 27.8%

Answers

Page 177: Repeated changes

1. a) £288 b) £558.90 c) £384.75
2. a) 3720 b) 3721
3. a) 420 b) 560
4. a) £3200 b) £2560 c) 36%
5. a) £1575 b) £1638 c) unchanged

Chapter 16: Probability

Page 181: Working out probability

1. a) $\frac{1}{4}$ b) $\frac{1}{2}$ c) $\frac{3}{13}$
   d) $\frac{1}{13}$ e) $\frac{1}{52}$
2. a) $\frac{1}{5}$ b) $\frac{4}{5}$ c) $\frac{2}{5}$ d) $\frac{2}{5}$
3. a) $\frac{1}{9}$ b) $\frac{1}{36}$ c) $\frac{1}{2}$
   d) 0 e) $\frac{1}{6}$

Page 183: Estimating probability

1. a) $\frac{23}{161}$ or 0.1429 b) $\frac{138}{161}$ c) 143
2. a) $\frac{1}{2}$ b) $\frac{15}{20} = \frac{3}{4}$
   c) it was probably wrong.
   d) experiment not conclusive.
3. a) $\frac{1}{200}$ b) $\frac{199}{200}$ c) 60
   d) (i) $\frac{24}{1200} = \frac{1}{50}$ (ii) $\frac{49}{50}$
   (iii) 240.
   e) Yes.

Index

A
Adding, negative numbers 30–31
Algebra 22–33
  brackets 28–29
  equations 120–129
  substituting into a formula 26–27
  writing 22–23
Angles 16–21
Approximations 130–143
  costs 134–135
  decimal places 130–131
  errors 140–141
  fractions 138–139
  percentages 138–139
  significant figures 132–133
  using calculators 136–137
Area 46–59
  circle 48–49
  parallelogram 46
  trapezia 46
Averages
  moving 72–73
  statistics 148–149
  4-point 72

B
Back-substitution 122
Bearings 158–159
Bills, money 170–171
Bivariate data 74–75
Box-and-whisker diagrams (boxplots) 116–117

C
Calculators 136–137
Capacity 10
Circles 3
  area 48–49
  circumference 48–49
Circumference, of a circle 48–49
Closed questions 153
Comparisons, fractions and percentages 42–43
Compound interest 166, 168–169
Continuous data, grouping 104–105

Costs, estimating 134–135
Cumulative frequency, 112–113
  curve 112
  polygon 112
  table 112
Curved graphs 88–89

D
Data 68–76
  bivariate 74–75
  continuous 104–105
  grouped 104–119
  mean 110–111
  median 110–111
  mode 110–111
  rounded 106–107
Decimal place, approximations 130–131
Decimals 4–19
Denominator 34
Dimensions 56
Discount, money 176–177
Distance 100–101
Dual bar chart 108

E
Equivalent fractions 34
Equations 120–129
  changing the subject 126–127
  of the graph 80
  solving equations 122–123
  with fractions 124–125
Errors 140–141
Estimating
  costs 134–135
  probability 182–183
Exact values 42
Expanding the brackets 28
Expression 24

F
Factors 4
Fractions 34–45, 138–139
  changing to percentages 38–39
  in equations 124–125
  simplest form (lowest terms) 34
  improper 34

mixed number 34
Frequency polygons 108–109
Frequency, cumulative 112–113

G
Gradient, graph 82–83
Graph 80–91
  curved 88–89
  gradient 82–83
  intercept 82–83
  travel graphs 86–87

H
Highest common factor (HCF) 12

I
Identities 123
Imperial units 10–11
Improper fractions 34
Independent events 180
Index forms 4
Intercepts, graph 82–83
Interest
  compound 168–169
  simple 166–167
Inter-quartile range

L
Length 3, 10, 156–165
Line of best fit, scatter diagrams 76–77
Loss, money 174–175
Lower quartile 114
Lowest common multiplier (LCM) 12

M
Mass 10
Mean 146
  grouped data 110–111
Median 146
  grouped data 110–111
Metric units 10–11
Mixed number fractions 34
Mode 146
  grouped data 110–11
Money 166–179
  bills 170–171
  compound interest 168–169

Index

discount 176–177
loss 174–175
profit 174–175
simple interest 166–167
value added tax (VAT) 172–173
Moving averages 72–73
Multiples 4

N

Negative numbers 62–63
adding 30–31
simplifying expressions 64–65
subtracting 30–31
Numbers
large 6
small 6
Numerator 34

O

Open question 153

P

Parallelograms 46–47
area 46
Percentage change 40
Percentages 34–45, 138–139
Pie charts 68–69
Pilot surveys 153
Polygons, frequency 108–109
Power 4
Primary data 153
Primes 4
Prime factorisation 12–13
Prisms 50–54
surface area 52–53
volume 50–51
Probability 180–185

estimating 182–183
Profit, money 174–175
Proportion 92–103
Pythagoras' rule 3, 160–163
Pythagorean triple 165

Q

Quadrilaterals 2, 18–19
Quartiles 114–115

R

Radius 54
Range 146
Ratio 92–103
changing money 98–99
distance, speed and time 100–101
unitary method 96–97
Reciprocal 34
Relative frequency 182
Reports 154–155
Rounded data, grouping 106–107

S

Scale drawings 156–157
Scatter diagram, line of best fit 76–77
Secondary data 153
Shapes 16–21
Significant figures, approximation 132–133
Simple interest 166–167
Speed 100–101
Standard form 6–7
Statistics 144–155
averages 148–149
making comparisons 150–151

reports 154–155
surveys 152–153
Stem–and–leaf diagrams 70–71
Subtracting, negative numbers 30–31
Surface area, of a prism 52–53
Surveys 152–153
Symbols 60–67

T

Terms, like and unlike 24
Time 100–101
Trapezia 46–47
area 46
Travel graphs 86–87
Triangles 2
Trigonometry 3, 160–163

U

Units 3
imperial 10–11
metric 10–11
Upper quartile 114

V

Variables 24
Variation, seasonal 72
Value added tax (VAT) 172–173
Volume 46–59
of a prism 50–51

W

Whole numbers 4–15

Y

y-intercept 84